PUBLIC TRANSPORT AND ITS USERS

Public Transport and its Users

The Passenger's Perspective in Planning and Customer Care

Edited by

MARTIN SCHIEFELBUSCH

and

HANS-LIUDGER DIENEL

nexus Institute for Cooperation Management and Interdisciplinary Research, Berlin

ASHGATE

Published by
Ashgate Publishing Limited
Wey Court East
Union Road
Farnham
Surrey GU9 7PT
England

Ashgate Publishing Company
Suite 420
101 Cherry Street
Burlington, VT 05401-4405
USA

www.ashgate.com

British Library Cataloguing in Publication Data
Public transport and its users : the passenger's
perspective in planning and customer care. - (Transport and
society)
1. Transportation - Planning - Citizen participation -
Cross-cultural studies 2. Transportation - Management -
Cross-cultural studies 3. Consumers - Attitudes -
Cross-cultural studies
I. Schiefelbusch, Martin II. Dienel, Hans-Liudger, 1961-
388

Library of Congress Cataloging-in-Publication Data
Schiefelbusch, Martin.
Public transport and its users : the passenger's perspective in planning and customer care / by Martin Schiefelbusch and Hans-Liudger Dienel.
p. cm.
Includes bibliographical references.
ISBN 978-0-7546-7447-4
1. Transportation--Environmental aspects--Europe--Case studies. 2. Consumer satisfaction--Europe--Case studies. I. Dienel, Hans-Liudger, 1961- II. Title.

HE147.65.S34 2009
388.4094--dc22

2008045783

ISBN 978-0-7546-7447-4

Mixed Sources
Product group from well-managed forests and other controlled sources
www.fsc.org Cert no. SA-COC-1565
© 1996 Forest Stewardship Council
FSC

Printed and bound in Great Britain by
MPG Books Ltd, Bodmin, Cornwall.

Contents

List of Figures

List of Tables

List of Abbreviations

AB	Aktiebolag
ABB	Allgemeine Beförderungsbedingungen
ACI	Automobile Club Italia
ADFC	Allgemeiner Deutscher Fahrrad Club
ADR	Alternative Dispute Resolution
AEG	Allgemeines Eisenbahngesetz
AFNOR	Association française de normalisation
AG	Aktiengesellschaft
AGB	Einheitliche Rechtsvorschriften für den Vertrag über die internationale Eisenbahnbeförderung von Gütern
AGIR	Association pour la Gestion Independante des Reseaux
AK	Arbeitskammer (Chamber of Labour)
AO, AOTU	Autorité organisatrice des transports urbains
ARN	Allmänna Reklamationsnämd
AVV	Aachener Verkehrsverbund GmbH
B	Belgium
BAB	Bus Appeals Body
BAG-SPNV	Bundesarbeitsgemeinschaft der Aufgabenträger des SPNV
BAV	Bundesamt für Verkehr
BB	Besondere Beförderungsbedingungen (special terms of carriage)
BEUC	Bureau Europeen des Union de Consommateurs
BGB	Bürgerliches Gesetzbuch (Civil Code)
BGBl.	Bundesgesetzblatt
BMELV	Bundesministerium für Ernährung, Landwirtschaft und Verbraucherschutz
BMVIT	Bundesministerium für Verkehr, Innovation und Technolgie (Österreich)
BR-Drs.	Bundesratsdrucksache
BSAG	Bremer Straßenbahn AG
BT-Drs.	Bundestagsdrucksache
BTTB	Bond van Trein- Tram en Busgebruikers
BUCT	Bus Users Complaints Tribunal
BVG	Berliner Verkehrsbetriebe AöR
BVO	Busverkehr Ostwestfalen GmbH
c.	circa
ČD	Česke dráhy

CER	Community of European Railways and Infrastructure Companies
CERTU	Centre d'Etudes sur les Réseaux, les Transports et l'Urbanisme
CH	Switzerland
CI	Consumers International
CIM	see ER CIM
CIPE	Comitato Interministeriale per la Programmazione Economica
CIV	see ER CIV
CL	Chamber of Labour (see AK)
CNCU	Consiglio nazionale consumatori e utenti
CNDP	Commission Nationale du Débat Public
CNEL	Consiglio Nazionale dell'Economia e del Lavoro
COTIF	Convention relative aux transports internationaux ferroviaires (Convention concerning International Carriage by Rail from 9 May 1980)
CPT	Confederation of Passenger Transport
CRUCC	Central Rail Users Consultative Committee
ČSAD	Ceskoslovenská automobilová doprava
CSS	Customer Satisfaction Survey
ct	cents
CVOV	Centrum Vernieuwing Openbaar Vervoer
CZK	Czech Krone
D	Germany
Dept	Department
DB AG	Deutsche Bahn AG
DfT	Department for Transport
DIN	Deutsches Institut für Normung e.V.
DPM	Direct Performance Measures
DTT	Direction des Transports Terrestres
EBG	Eisenbahnbeförderungsgesetz (Österreich)
EC	European Community
EDS	extr-ajudicial dispute settlement
EGV	Vertrag zur Gründung der Europäischen Gemeinschaft (Treaty for the Foundation of the European Community)
ENeuOG	Eisenbahnneuordnungsgesetz
EORG	European Opinion Research Group
EPCI	Établissement public de coopération intercommunale
EPF	European Passengers' Federation
ER CIM	Einheitliche Rechtsvorschriften für den Vertrag über die internationale Eisenbahnbeförderung von Gütern

ER CIV	Einheitliche Rechtsvorschriften für den Vertrag über die internationale Eisenbahnbeförderung von Personen und Gepäck
ES	European Standards
EU	European Union
EUSG	EU Service Guarantees
EVO	Eisenbahn-Verkehrsordnung
e.V.	eingetragener Verein
EVU	Eisenbahnverkehrsunternehmen
F	France
FGSV	Forschungsgesellschaft für Straßen- und Verkehrswesen
FNAUT	Fédération Nationale des Associations d'Usagers des Transports
FoE	Friends of the Earth
FS	Ferrovie dello Stato
GART	Groupement des autorités responsables de transport
GB	United Kingdom of Great Britain and Northern Ireland
GDP	gross domestic product
GDR	German Democratic Republic
GEB	Community of European Railway and Infrastructure Companies
GmbH	Gesellschaft mit beschränkter Haftung
GPR	Zeitschrift für Gemeinschaftsprivatrecht
GTC	General Terms of Carriage
HGB	Handelsgesetzbuch (Code of Commerce)
HVG	Heidenheimer Verkehrsgesellschaft mbH
IC	InterCity (DB AG)
ICE	InterCityExpress (DB AG)
IGEB	Interessengemeinschaft Eisenbahn, Nahverkehr und Fahrgastbelange Berlin e.V.
IPFAS	Independent Penalty Fares Appeals Service
ISFORT	Istituto Superiore di Ricerca e Formazione per i Trasporti
ISO	International Standards Organisation
JHMD	Jindřichohradecké místní dráhy
KflG	Kraftfahrliniengesetz (Österreich)
Klausel-RiLi	EG-directives 93/13/EWG from the Council of 5 April 1993 regarding improper clauses in consumer contracts
KNV	Koninklijk Nederlands Vervoer

KOV	Konsumentverket, Sweden
KpVV	Kennisplatform Verkeer en Vervoer
KVB	Kölner Verkehrsbetriebe
LA –TC	Lower Austrian Transport Concept (see NÖ-LVK)
LBO	Landesverband Bayerischer Omnibusunternehmen
LEADER	Liaison entre actions de développement de l´économie rurale (relationship between actions for the development of the rural economy)
LOCOV	Landelijk Overleg Consumentenbelangen Openbaar Vervoer
LOTI	Loi d'orientation des transport intérieurs
LPT	local public transport
LRPC	London Regional Passengers' Committee
LSA	Lichtsignalanlage
LPRT	local passenger rail transport
LTP	Local Transport Plan
LTUC	London Transport Users' Committee
LTW	London TravelWatch
LuftVG	Luftverkehrsgesetz
LVB	Leipziger Verkehrsbetriebe
MA	Montréal Accord on the standardisation of particular provisions regarding carriage in international air transport
MET	Metropolitan (DB AG)
MID	Mobilität in Deutschland
MIT	motorised individual transport
MOBIS	Branchenvereinigung im ÖPNV der Niederlande
MSS	mystery shopping surveys
MüKo	Münchener Kommentar zum Bürgerlichen Gesetzbuch
NJW	Neue juristische Wochenschrift
NL	The Netherlands
NÖ-LVK	Niederösterreichisches Landesverkehrskonzept
NÖVOG	Niederösterreichische Verkehrsorganisationsgesellschaft
NPS	National Passenger Survey
NRW	North Rhine-Westphalia
NS	Nederlandse Spoorwegen
n.s.	not stated
NVG	Niederschlesische Verkehrsgesellschaft mbH
NVV	North Hessian VerkehrsVerbund
NZV	Neue Zeitschrift für Verkehrsrecht

ÖBB	Österreichische Bundesbahn
OÖVV	Oberösterreichische Verkehrsverbundgesellschaft, Linz (see UATNA)
OPC	Ondersteuning collectief Personenvervoer aan Consumentenorganisaties
ÖPNRV-G	Gesetz zur Ordnung des öffentlichen Personennah- und Regionalverkehrs
OV	Openbaar vervoer
P+R	Park and ride
PBefG	Personenbeförderungsgesetz
PF	Passenger Focus
PLM	Passenger Link Manager
PPC	Public Procurement Code
PPT	public passenger transport
PT	public transport
PTA	Public Transport Authority
PTE	Passenger Transport Executive
PTU	Périmètre de transports urbains
PTS	Programma Triennale Servizi
RATP	Régie autonome des transports Parisiens
RegG	Regionalisierungsgesetz
RDS	Railway Development Society
RET	Rotterdamse Elektrische Tramweg
RFI	Rete Ferroviaria Italiana
ROCOV	reizigersplatforms in het regionaal openbaar vervoer
ROVER	Reizigers Openbaar Vervoer
RPC	Rail Passenger Council
RRa	Reiserecht aktuell
RTC	rail transport company
RUCC	Rail Users' Consultive Committee
S	Sweden
SCG	Schienen Control GmbH (Österreich)
SGC	Stichting Geschillencomissies voor Consumentenzaken
SD ČR	Svaz dopravy České republiky
SDP	Sdružení dopravních podniků České republiky
SEK	Schwedische Krone
SHA	Stadtbus Schwäbisch Hall GmbH
SIKA	Statens Institut för Kommunikatinons Analys, Schweden
SJ	Statens Järnvägar
SL	Storstockholms Lokaltrafik, Schweden
SLTF	Svenska Lokaltrafikföreningen

SNCB	Société Nationale des Chemins de fer Belges
SNCF	Société Nationale des Chemins de fer Français
SOU	Statens Offentliga Utredningar
SRL	Vereinigung für Stadt-, Regional- und Landesplanung e.V.
SSB	Stuttgarter Straßenbahnen AG
STIB	Société des Transports Intercommunaux de Bruxelles
STIF	Syndicat des Transports des Ile-de-France
SYTRAL	Syndicat mixte des transports pour le Rhone et l'Agglomération Lyonnaise
TA	Technikfolgenabschätzung
TARP	Technical Assistance Research Program
TC	transport company
TGV	Train à Grande Vitesse
TNOA	Transport Network Organisational Organisations (see VVOG)
TOC	Terms of Carriage
TÖB	Träger öffentlicher Belange
TranspR	Transportrecht
TUCC	Transport Users' Consultive Committee
UATNA	Upper Austrian Transport Network Association, Linz (see OÖVG)
UITP	International Association of Public Transport
ÜSTRA	Überlandwerke und Straßenbahnen Hannover AG
UTP (F)	Union des transports publics
UTP (I)	Associazione Utenti del Trasporto Pubblico
VAG	Freiburger Verkehrs AG
VBB	Verkehrsverbund Berlin Brandenburg
VBN	Verkehrsverbund Bremen/Niedersachsen GmbH
VCD	Verkehrsclub Deutschland e.V.
VCÖ	Verkehrsclub Österreich
VDV	Verband Deutscher Verkehrsunternehmen
VgV	Verordnung über die Vergabe öffentlicher Aufträge
VOL/A	Verdingungsordnung für Leistungen
VOR	Verkehrsverbund Ostregion, Wien.
VöV	Verband öffentlicher Verkehr
VRR	Verkehrsverbund Rhein-Ruhr AöR
VRS	Verkehrsverbund Rhein-Sieg GmbH
VVOG	Verkehrverbundorganisationsgesellschaft
VVS	Verkehrs- und Tarifverbund Stuttgart GmbH
vzbv	Verbraucherzentrale Bundesverband

Wp	Wet personenvervoer
WVG	Westfälische Verkehrsgesellschaft mbH

About the Authors

Dr **Dipl.-Ing. Hans-Liudger Dienel** was born in Münster, Germany in 1961. He is a founder and director of the nexus Institute and since November 1995, academic director of the Center for Technology and Society at the University of Technology (TU), Berlin. Dr Dienel studied history, philosophy, and engineering from 1983 to 1990 at TU Munich and completed a doctorate there in 1993 about the interaction between academia and industry in technical sciences in Germany and the USA. From 1989 to 1995 he was a research associate at the German Museum Munich where he lead the task group on transport history from 1993 to 1995. The focuses of his research include future studies, participation in urban and technological development, mobility research, and transport policy. In the field of transport, Dr. Dienel has published, among other things, numerous works about transport history, mobility for older people, and leisure transport.

Dr **Martin Schiefelbusch** MA MSc was born in Kassel, Germany in 1970. From 1992 to 1998 he studied transport planning and geography at the University of Technology (TU) Berlin, punctuated by a one-year stay in London during which he completed the MSc programme of study 'Transport Planning and Management' at the University of Westminster. In March 2007, he completed his doctoral studies at TU Berlin. During his studies he also worked in several positions in the public transport field in service planning, research, and interest representation. After working at Prignitz Railways and the Center for Technology and Society at TU Berlin, since 2000 Martin Schiefelbusch has been a research associate at the nexus Institute and since 2007 has been head of the Mobility/Space/Demography division. The focuses of his work include service development for PT, participatory procedures, implementation strategies, and transport policy issues. He has published several contributions to these fields.

Dipl.-Ing. **Christian Bodensteiner** was born in 1975 in Neustadt on the Waldnaab, Germany and studied urban and regional planning at the University of Technology (TU) Berlin and at the University College Dublin. Since 2005 he has been a freelance associate at the nexus Institute Berlin focusing particularly on issues of local transport planning and procedures for customer participation in transport. For the nexus Institute he has researched, among other things, various participatory procedures for transport planning and the state of passengers' rights in the EU. In addition to his work with the nexus Institute, he works as an urban and transport planer in Berlin.

Dipl.-Ing. **Uwe Böhme**, born in Leipzig in 1977, studied transportation at the University of Technology (TU) Berlin and has been a freelance associate at the nexus Institute since 2005. His primary focus is issues of public passenger transport, particularly matters of transport service marketing and complaint management. He has also carried out comparative analyses of passengers' rights and participation. In addition to his scholarly work for the nexus Institute, Uwe Böhme works as a transport engineer for a planning consultancy in Berlin and as a freelance associate for other projects related to transport planning.

John Cartledge is the Deputy Chief Executive of London TravelWatch, the official organization of passengers in the London metropolitan area (further information at www.LondonTravelwatch.org.uk). John Cartledge has been working for the organization since 1975 and since 1987 has been focused on transport policy issues especially relating to safety. In addition, he advises a number of organizations and research institutes in the British rail transport market.

Dipl.-Ing. **Hans-Peter Faas** was born in 1951 and studied civil engineering at the University of Stuttgart with a focus on the field of transport. After working in managerial positions in planning consultancy (Obermeyer in Munich, Frebner-Ruchay in Berlin), hes was the Director of Market Research from 2004 to 2007 at the Berliner Verkehrsbetrieben (BVG).

Trevor Garrod, born in 1946, has been the Chairman of the European Passengers' Federation since 2002. Since the beginning of the 1970s, he has been active in passengers' associations in Great Britain, most recently at the association 'Railfuture'. He lives in Lowestoft, a town about 200 km northeast of London. For 24 years, he has been commuting to work on the train. He previously worked as a teacher, and for his current position in the public administration he can reach his workplace on foot.

Dipl. Soz.-Wiss. **Holger Jansen** was born in 1968 in Monheim (Rhein) in Germany. He studied political science, law, and geography in Duisburg. Parallel to his studies, he was engaged as a volunteer for the federal passengers' association Pro Bahn, acting a spokesperson (1994–2000) and as Assistant Chairman (1996–2000). After completing his studies, he was a founding member of the Allianz pro Schiene (2000–2001). Afterwards, he directed a project for passenger education about passengers' rights at the Federal Association of Consumer Organizations. Since 2004, he has overseen several projects at the nexus Institute in the fields of mobility and development of rural regions in East Germany.

Dipl. Ing. **Hartwig Meier**, born in 1963, studied history, chemistry, and spatial planning at the Universities of Bielefeld and Dortmund. Since 2001, he has been the Head of Transport Planning for moBiel GmbH, the transport company of the Bielefeld municipal utilities. In the same company, he was the project director for the institution of a quality management system. Starting in 2003, he has been working in the German and European task groups for the formulation of the EN 15140.

Dipl. Geogr. **Nancy Neugebauer**, born in 1978 in Berlin, began studying to be a travel agent and completed her studies in geography, sociology, and business management at Humboldt University in Berlin. During her studies she worked for, among other things, a cruise agency in Baden-Württemberg and for the Verkehrsclub Deutschland (VCD), National Association Berlin. Since 2006, she has been a research associate at the nexus Institute and works in the areas of service marketing, customer service, and passengers' rights in public transport.

Dr **Mihael Aleksander Pohar** was born in Bruchsal, Germany in 1977 and grew up in the Graf Berg Region near Gummersbach. In 1996 after his Abitur (German university entrance diploma) and social service work in Bergneustadt in the Oberbergian district, he studied law at the Westphalian Wilhelm University Munster and finished his studies in 2002 with distinction. Until 2006, he worked as a research associate at the Institute for International Business Law chaired by Prof. Dr Heinrich Dörner. From the start there, he became interested in the field of passengers' rights and published contributions to this field and others, particularly articles related to travel law in legal professional journals and consumer journals. His dissertation 'Rechtsbeziehungen zwischen Fahrgast und Eisenbahn' ('Legal Relationships between Passenger and Railway'), published in autumn 2006, was recognized with the sponsorship prize of the German Society for Travel Law. In 2007, he was appointed to be a judge in Düsseldorf.

Anthony D. Smale was born in Surrey, England in 1947. Since 2000 he has been a Transport Officer in the Hampshire County Council and responsible for the formulation of transport contracts in LPT and the organization of student bus transport. He is a member of the Association of Transport Coordinating Officers. For over 20 years, Anthony D. Smale has been an active member of Railfuture (previously the Railway Development Society) and since 2002 he has been Railfuture's Rail User Group Liaison Officer.

Dipl.-Ing. **Jenny Schmithals** MA studied environmental technologies and sociology at the University of Technology (TU) Berlin and the Technical University of Denmark/Lyngby and specialized in the sociology of mobility, social ecology, and demography. In Berlin she worked on specific transport political issues such as the reorganization of public transport in urban and airport policy. Since 2002, she has been an associate of the nexus Institute for Cooperation Management and Interdisciplinary Research where she works on inter- and transdisciplinary research projects and in the fields of cooperation management and moderation.

Dr **Huibert A. Verdoold**, born in 1973, studied management sciences at the Erasmus University Rotterdam. Since 2002 he has been working as a consultant for OPC (www.stichting-opc.nl). He assists ROCOVs in the organization and design of user-participatory procedures. He is an expert advisor for the introduction of a chip card for the entire public transport system in The Netherlands, which was scheduled to be introduced 1 January 2009.

Preface

Our work on this book has been supported by numerous colleagues over the course of five years. It certainly does not mark the end of the discussion, but the conclusion of one of its stages. We would like to take this opportunity to thank all of the people who have helped us in our efforts to understand and insightfully advance the development of passengers' rights and passenger participation.

This includes, first of all, the Volvo Foundation in Gothenburg/Sweden, without whose generous support this project could not have been realized. It is an exceptional international institution that gives not only financial support, but also offers a platform for the integration of mobility research. In 2005, the Foundation gave us the opportunity to conduct an international conference on the topic 'More Customers through More Participation?', which greatly enriched our work.

We would especially like to thank the authors both inside and outside of our Institute, as well as our numerous colleagues in ministries, associations, companies, advisory councils and other institutions that have shared their experiences with us. We would particularly like to name Hartwig Meier, Mihael Pohar, Hans-Peter Faas, Huibert Verdoold, Anthony Smale, Trevor Garrod and John Cartledge, as well as the nexus Institute members Holger Jansen, Nancy Neugebauer, Christian Bodensteiner, Uwe Böhme and Jenny Schmithals. We also thank Štěpánka Doleželová and Patrick André for their contributions to the country reports. For their energetic and patient support in transforming our work into this book, we would like to thank Ashgate Publishing, our translator Eric Green as well as Stephan Feldhaus, Patrick Ebers and Tom Tischendorf for editorial work. Chris Cook in London provided consultation for the revision of the manuscript for English-speaking readers.

The European Commission and the German Federal Ministry of Consumer Protection allowed us to explore many international issues more deeply through research contracts. We also extend heartfelt thanks for support from Peter Faross, Hein Bollens and Marie-Luise Altmutter (all EU Commission), Jürgen Karwelat (BMELV) as well as our project partners Alexander Schulz and Dennis Tänzler (both Adelphi-Research), Ola Svenonius, Stefan Albrecht, Olivia van Riesen, Louis Kaeppelin, Efthymis Zagorianakos, Sara Björlin-Lidén, Annalisa Giovani and Stefano Maggi.

Martin Schiefelbusch and Hans-Liudger Dienel
Berlin, Easter 2008

Introduction

Hans-Liudger Dienel

Do better passengers' rights strengthen customer satisfaction and lead to more passengers, and ultimately to growth and improvement in public transport? Or do enforceable passengers' rights weaken already loss-making transport companies? This book gives clear, documented answers to these questions central to the future and the potential of public transport: In almost all cases that passengers' rights have been codified and enforced, the number of passengers has increased, the image of public transport and of transport companies has improved; and despite additional expenses, the economic situation of the transport industry has improved as well. A discerning view shows that, particularly in long-distance transport, extensive reimbursement regulations have proved of value; while local transport still has little experience with major compensatory services.

This book builds on around five years of European comparative research by the nexus Institute about the customer's position in public transport. In the context of Europe, it comparatively documents the widespread rise of passengers' rights in several countries, as well as the direction of the path before us, which in the meantime has started to seem irreversible. According to the authors' predictions, it will lead to a broader strengthening of public transport in Europe.

In recent years, the most important motor for the strengthening of passengers' rights in Europe has been the European Commission, thus strikingly demonstrating its citizen-orientation. The EU Commission closely observes individual improvements on the national level and uses them as arguments for the development of new supranational instruments.

The second topic of this book is passenger participation in service design. There is a connection, if only indirect, between passengers' rights and passenger participation. Both areas address a new appreciation of the passenger: the identification of the passenger as a customer with enforceable rights instead of merely something that is transported, and the inclusion of passengers as collaborative counterparts and decision-makers in identifying weak points and developing proposals for the improvement of the public transport system.[1] *Passenger participation and passengers' rights have significant overlapping areas, and both have a common overall goal of strengthening public transport.*

1 A good overview of participatory practices in transport is provided in the report of the EU project GUIDEMAPS (Gaining Understanding of Improved Decision Making and Participating Strategies). A handbook was also produced (Kelly et al. 2005), which can be ordered free of charge.

Both approaches are still relatively new in public transport. In its combination of passengers' rights and passenger participation, this book is the first substantial work about this topic. It is truly astonishing that an issue so important for the future of public transport has to date been so neglected.

This book approaches the topics of passengers' rights and passenger participation from not only a comparative transport political perspective, but from judicial and socio-political perspectives as well. The transport political dimension has already been addressed. The judicial perspective particularly emphasizes the necessity of enforcing passengers' rights as customers in relation to their contractual partner, the transport company. This perspective is important and fully justified, but not without its own dangers. *Justitia non calculat* (jurisprudence does no accounting). This old legal motto could be radically applied – in local transport for instance – to passengers' rights, which many transport companies fear will undermine their basic economic conditions. They refer to legal consumer protection in the United States, which has endangered some commercial sectors due to indemnity claims that are sometimes exorbitant. This certainly has a tradition in the USA. In the first decade of the twentieth century, railways were so weakened by cartel bans and law suits that they almost had to completely give up their fight for passenger transport against competition on the streets. In Europe, the fears that a strengthening of passengers' rights could weaken the position of transport companies is, however, not empirically demonstrable.

From a socio-political perspective, it is overall desirable to strengthen civic involvement in public commodities. Public transport is a public commodity. Just as the judicial perspective places passengers' rights at the centre, the socio-political perspective emphasizes passenger participation.

Passenger participation is about the development and adaptation of new, more customer-oriented transport systems and the development of solutions that often occur more easily to the customers than to the experts. Customers are experts, so to speak, in their own affairs – and this is not limited to the design of seats, etc. Customer participation is becoming increasingly important for the development of new and better services in many sectors. In the 1950s, it was often sufficient for the planning and development of new transport services if transport engineers and planners acted on even their own instincts. Today, technology developers and transport planners have to consult their customers, because post-industrial society has become more starkly divergent, and because engineers and planners themselves often no longer reflect the mainstream of society. The technology historian and innovation researcher Ulrich Wengenroth[2] calls them the 'farmers of the twenty-first century'. We need farmers in industrial society as well, but they are no longer the core of the society.

2 Wengenroth (2001).

The inclusion of passenger associations is also part of customer participation. It is a sad situation that the over a hundred passenger associations in German-speaking countries alone often have a latently tense relationship with transport companies. What should be aspired to, in contrast, is a relationship like that between a supporters' club and the management of a well-lead football organization.

Chapter 1 provides a brief introduction to the topic of 'passenger interests'. The book then proceeds in four main parts: two main sections (Chapters 2 and 4) are focused on passengers' rights and passenger participation, complimented by a section (Chapter 3) on quality management, and a section (Chapter 5) on European comparison. A short sixth chapter provides conclusions and perspectives for further development.

The aim of this book is to provide both analysis and advice. Hence each chapter can be read on its own, and due to the fact that there are links between different issues in practice, some are addressed in more than one location. As the present version of this book is based on a similar text for a German readership, the German situation is used to illustrate some of the issues raised in the following chapters. However, case studies and good practice examples from around Europe are included in Chapters 2 to 4 as well.

Chapter 1
Passenger Interests in Public Transport

Martin Schiefelbusch

After the introduction, and before the different possibilities for participation and representation of user interests are discussed, this chapter aims to provide a brief overview about the subject of this book. The following pages therefore describe the particularities of consumer representation in public transport and what distinguishes participation in this area from that in other areas of planning.

The chapter is divided in two parts: Section 1.1 presents the institutional setup under which public transport services operate. Section 1.2 goes on to define customer interests in this field – what do they expect in the different stages of planning and providing a transport service?

1.1 The Institutional Context of Public Transport Provision

1.1.1 Dominating Monopolies

The mass provision of public transport (PT) services in cities started as a private sector industry in the nineteenth century, although in most parts of Europe as well as in other developed countries these days are long gone.

Urban and regional public transport, where the lion's share of journeys are made, has for decades usually been provided either by public sector companies or by operators working under licences granted by public authorities and hence under monopoly conditions. As a result, typically only one service provider is available to satisfy a particular transport need fixed by route, time and additional service requirements.

For long-distance transport the situation is slightly different as competition between different modes – for example, railways, long-distance coaches and aviation – has been less strictly regulated, and liberalization started earlier here than it did in local markets. Hence competition between different providers can exist in a particular corridor, but even here it is not necessarily intense. Furthermore, many of the private operators that existed in the nineteenth and early twentieth century have been transformed into publicly-owned companies which operate across the country. Despite the more or less comprehensive liberalization of air, coach and rail markets in the recent past, these providers usually still dominate their national markets.

In this context one ought to remember that this market structure is not restricted to regulated markets or nationalized operators: Even in the UK, where public transport was deregulated in the 1980s,[1] on-street competition has largely ceased and is now limited to a small number of very busy corridors in the main metropolitan areas. Otherwise, local services have reverted to local monopolies which – in contrast to the old system – now are almost exclusively privately owned.[2] On the privatized British rail market, competition outside the system of franchised operations has been slow to develop as well.

Consequently, customers cannot express dissatisfaction with the service on offer by switching to another operator, the so-called 'exit' option.[3] Therefore a key incentive for the provider to orientate its service to the users' needs is missing.

The only alternative for 'exit' is to choose another mode, as demonstrated by the reducing market shares and/or passenger numbers of public transport in the last decades. However, to choose other modes is not always practical and usually – in case of car use – not desirable from the perspective of transport policy.

According to Hirschman, the alternative to 'exit' gives users a 'voice'. As will be discussed later, possibilities for doing so are at present not well developed either in many regions, hence to 'voice' criticism is also fraught with difficulties.[4] But there are ways in which transport providers, authorities and civil society organizations can improve the position of passengers. These possibilities are therefore the main focus of this book, leaving aside for the moment thoughts about better 'exit' – that is competition on the market.

1.1.2 Political Intervention

Political influence on service planning, provision and funding has always been strong because cities, regions and states are either owners of transport companies or influence the service through regulation or contracting policy. They decide on the amount of money available to run services and maintain infrastructure and also shape intermodal competition with political decisions.

As a result, passengers might seek to influence service provision by means of the political process either through lobbying activities or by making use of political authorities' role as the owners of service providers. However, this approach is indirect and time-consuming. The chances of being heard are bound by the stakeholders' position in the political system where they have

1 Local public transport outside London and Northern Ireland, taxi services (deregulated in 1986) and interurban coach services (1980).

2 Cf. Anon. (1997), Simpson (1996), pp. 216–18, White (1995).

3 Hirschman (1970).

4 Hanson (1999), p. 38, Cartledge (2005), Schiefelbusch (2005a).

to compete for attention and resources with other policy objectives. Without suitable procedures, these circumstances are likely to have a negative impact on the users' possibilities to 'voice' their concerns.

Regulatory reform that has taken place on the national and European level since the 1990s has added another dimension to this dilemma. Following the EU's concept of market liberalization, subsidies can only be paid to operators after non-discriminatory tendering procedures or based on contracts between providers and authorities. This means that the tendering/contracting authority becomes the operator's most important business partner, as deficit compensation typically represents the largest individual source of revenue. Income from fares, on the other hand, has to be generated from a far greater number of individual users who are much more difficult to win over.

1.1.3 A Special Legal Relationship

In contrast to most other sectors, public transport operators have traditionally been exempted from civil law obligations requiring producers to replace or compensate consumers for the malfunctioning of their products. It is thus at the operator's discretion whether they offer compensation to passengers, a situation which in the past has provided little incentive for good customer care or complaints handling. As a result, the legal situation can be described as follows: the ticket purchased does not guarantee the customer that a certain journey can be undertaken based on the published information but rather offers the 'permission to use – within the spatial and temporal limitations of the ticket – a public transport vehicle if one happens to pass by'.[5]

This situation clearly puts the users at a disadvantage, both directly for those affected and indirectly because it can be interpreted as another incentive to reduce service standards. A controversial debate about the adequacy of these lower consumer care standards has developed in recent years (see Sections 1.2.4, 2.6, 2.7).

1.1.4 The Role of Public Transport for Society

Transport has traditionally been seen as not just an ordinary commodity, but also a prerequisite for many other economic and social activities. The role of public transport has changed over time. Instead of providing the backbone for the movement of passengers and goods, as in the nineteenth and early twentieth century, its role today is to be available as an alternative to the private automobile. Such an alternative is desirable for several reasons, such as environmental, energy and social policy objectives. But in order to succeed, its hybrid position between market and politics has to be borne in

5 As expressed by Rainer Engel, Pro Bahn, on the conference 'Derailed – Out of Track', Berlin, 5 June 2002, referring to the German situation.

mind. On the one hand, the political sphere shapes PT's competitive position, as explained elsewhere in this section, in particular by providing funding to address those objectives which would be left out if operators were to adopt a purely commercial perspective. On the other hand, however, despite these different interests, in the end public transport has to prove itself on the transport *market* in competition with other modes, because in most situations modal choice cannot be influenced by prohibitions or obligations.

This hybrid nature also creates two distinct leitmotifs for transport user representation. It can be driven by the experiences they have as consumers, but also – and perhaps even more so – by the conviction to support other objectives if public transport becomes more successful (Section 4.6.1). The work of passenger organizations therefore has similarities to that of other pressure groups campaigning for nature protection and other environmental aims.

The posited starting point for the considerations presented in this book is that the market structure described above does not guarantee that users' views will be integrated into the planning process. Hence there is a need for practices and structures that allow public transport passengers to express their interests directly, efficiently and in a timely manner.

1.2 Dimensions of User Interests

The reference to customer care might lead one to consider the position of public transport users as a problem for those departments dealing with complaints, comments and direct interaction between the provider and its customers. However, the previous section has shown that the passengers' position is also influenced by the wider framework. An effective dialogue with the customer must therefore go beyond the typical activities of customer care.

Looking more closely at the issue of user interests, it becomes apparent that these cannot be reduced to ex-post compensation for a lack of quality alone. An easy resolution of conflicts on the personal/individual level is a poor compensation if the service as a whole does not meet the users' needs. Hence a comprehensive view on the whole process of service planning and production is necessary. In doing so, four levels or layers of user interests can be identified which will be discussed in more detail in the following sections:

- on the *political level*, the competitive framework for public transport is set up and the strategic decisions on the service level are made;
- on the *planning level*, concepts for the service are developed and planned in detail including the preparation of operations and the level of spare capacity provided;

- on the *provision level*, these concepts are implemented, one of the main objectives being that deviations from the pre-planned pattern are kept to a minimum;
- on the *practical level*, however, problems cannot be avoided and therefore one needs to think about practical solutions to problems arising from deficits in quality.

The various issues which have to be addressed are of course connected and not always limited to just one of these levels. Similarly, different approaches can be developed for transport users and their organizations to express their views.

Figure 1.1 demonstrates these interrelations. The main issues are shown in the upper part, the different instruments in the lower. Both are arranged in relation to each other and to the four analytical levels described before. Different shadings indicate full or partial relevance of the individual aspects. This structure also provides a guideline for the following chapters. Chapters 2 to 5 discuss approaches for participation and representation of passenger interests in planning, quality management and customer care, complemented by summaries of the situation in different countries. First however, the various levels – the upper part of Figure 1.1 – will be presented to illustrate and structure the range of issues within this field. Due to its strategic importance, the political level will first be discussed in detail.

1.2.1 Politics: Maintaining and Improving the Position of Public Transport

On the political or strategic level, users' interests are directed at topics such as parking schemes, the decision to introduce or withdraw urban rail services and also land-use guidelines and tax incentives. All these can be combined in transport master plans or similar documents. Decisions at this level have direct and indirect impacts on the balance between modes and the conditions under which public transport services are provided. The general aim is to create an environment that favours public transport, in particular by giving it due consideration in the development of an urban transport strategy and the setting of political priorities.

There are several characteristics of transport policy that make this task a difficult one:

- due to the role of mobility as a prerequisite for other activities, it has numerous, albeit not always easily detectable, links with other policy fields;
- conversely, transport has multiple impacts on people and environment which are often seen as negative. Those not using the particular mode or service themselves are often particularly sensitive to these external

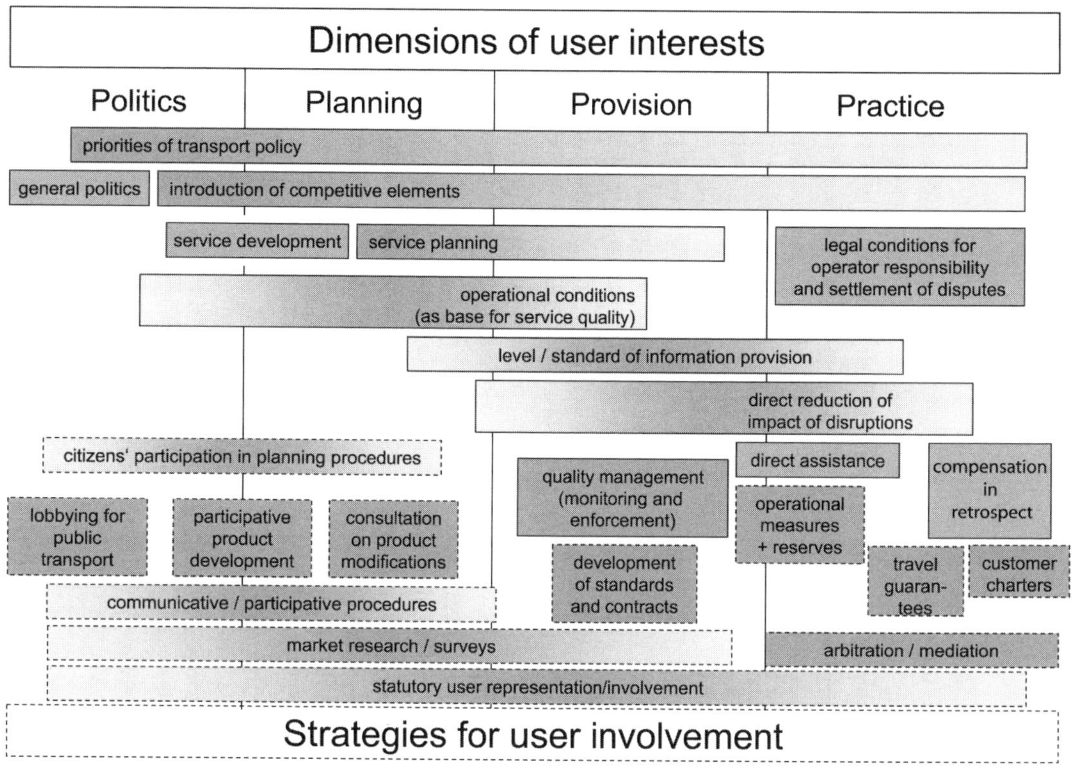

Figure 1.1 **Topics and approaches for user participation in public transport**

Source: author's depiction.

effects. The associated costs differ greatly between modes and are internalized to different degrees;

- the act of travelling itself is not necessarily seen as stressful. Rather, it can – within certain limits – even be perceived as an activity that serves psychological, physiological and social needs. It is usually easier to satisfy these with individual means of transport than by using public transport services, giving the former an additional appeal;[6]
- transport planning, however, tends to underestimate the role of these latter factors because it focuses on the technical and operational design of the transport system.[7] Furthermore, during the last few decades the focus of interest has often been placed on catering to motorized individual transport first. Today, the automobile-oriented city has largely ceased to be a model for planning, but the dominant position of the private automobile has had deep impacts on space and society.[8]

What does this mean for passengers' interests in transport policy? First one has to acknowledge that bus and rail services have rarely been at the centre of the policy debate in the past, even though there also have been exceptions. The ascent of the private automobile found widespread support especially from the 1950s to early 1970s, among politicians and planners, but also in the wider public (Table 1.1). Following the oil crisis and environmental disasters, this position was reverted to some extent, and until the late 1980s people looked more favourably at supporting public transport and even restricting car use. Since the mid-1990s, environmental concerns and willingness to accept traffic restrictions have lessened again, but not reverted to the previous position. The role of PT is thus more readily acknowledged, but growing attention to efficiency and effectiveness have lead to new arguments to be addressed.

At this stage, one should also bear in mind that *statements and perceptions do not always correspond with the reality* of the roles filled by the different modes of transport. Public statements about priority for public transport are frequently made. This does not, however, necessarily mean that the old view that buses and trains are 'really something just for the captives' (poor, aged, schoolchildren and so on) has disappeared completely.

However, this traditional perception of the different modes' importance does not correspond to reality either – at least not as an absolute rule. The impression of automobile traffic as the dominant mode already starts to change once all travel options are considered – including walking and cycling – instead of the 'classic' but incomplete modal split between cars and public transport. Secondly, the share of PT varies greatly according to the situation. Thirdly, if suitable services are provided, passengers tend to come in increasing

6 See, for instance, Klühspies (1999, 2001), Hilgers (1994), Dick (2002).

7 Schiefelbusch (2007b).

8 See, for instance, Reiter (1999), Schmucki (2001), Lundsgaard-Hansen (2000).

Table 1.1 Mission statements in the phases of transport planning

Phase	Approx. duration	Characteristics and objectives
Reconstruction	1945–55	Repair of damage from the war without redesigning, PT expected to continue bearing the brunt, but nevertheless planning for car traffic
Growth	1955–75	'Car-friendly city' as a reaction to the 'transport emergency', general acceptance and support of MIT, separate planning of PT and MIT, 'planning euphoria'
Downtrend	1975–90	Attempt to limit car traffic growth and mitigate its negative effects, initial measures including PT construction, then inclusion of non-motorized transport and land-use planning
Economizing	1990–2000	Shortage of public funds, streamlining, liberalization, use of capital reserves through improved governance (transport system management)
Consolidation	Since 2000	Cooperation between transport modes with a common goal of sustainability

Note: The times refer to Central Europe, The individual phases can differ in time and intensity in the various countries according to the development of motorization and environmental consciousness.

Source: author's compilation based on Reiter 1999, Schmucki 2001 and Lundsgaard-Hansen et al. 2000.

numbers from a wide range of social and economic backgrounds, in particular if occasional users are considered as well.[9] Lastly, decision makers tend to perceive the situation as more automobile-dependent and automobile-oriented than the general public does.[10]

Arguments for public and non-motorized travel options should also bear in mind the emotional dimension of mobility mentioned above. Transport policy discourses tend to focus on the concrete and rational advantages and disadvantages of alternative solutions, such as ecological problems caused by car traffic or safety gains through considerate driving. Environmental pressure groups are particularly prone to choose such arguments, but passenger organizations and public transport marketing campaigns do so as well. The

9 The last nationwide travel survey in Germany showed that about 40 per cent of the population used buses and trains at least once a month (data for 2002, cf. infas and DIW (2004), p. 42)

10 See the surveys undertaken by Brög in different European cities in the 1980s and 1990s (for example Brög (2000), pp. 3, 19–20).

problem is that this easily causes resentment in the targeted audience and thus the opposite of the intended result.[11]

This may well play more of a role for public relations activities than the dialogue with professionals, but should nevertheless not be ignored. Politicians aim to keep the support of their electorate and thus need to think about how their actions are perceived. Furthermore, at least on the local level, their mandate is based on part-time volunteer commitment and their own views therefore close to a lay persons' perspective.

1.2.2 Planning: Translating Passengers' Expectations into Services

On the planning level, public transport services are proposed and designed in detail. The topics which fall under this heading are manifold in terms of their scope, geographic coverage, and the degree of innovation required to assess the problem. Typical issues include the routing of services, timetables, ticket conditions, design of vehicles and location of stops, but also the development of new service concepts.[12]

In reality, the two levels naturally overlap: On the one hand, planning matters may become political ones even on a very small scale if users or other stakeholders cannot achieve a solution that satisfies them. On the other hand, political stakeholders may also wish to prescribe small details of the service pattern if they consider this important for their strategy. NIMBY[13] protests against new bus routes and 'political' fare reductions are examples of this. The close links between operators and (local) political authorities as their owners – so far common in many European countries – has certainly favoured such influences.[14]

A practical distinction between the political and planning level can be based on who is involved and who makes the decisions. Issues at the strategic level are normally decided by the legislature while on the planning level decisions mainly involve transport operators and the local administration.

Planning involves dealing with simple questions as well as complex assessment and decision processes. Furthermore, there is a wide range of stakeholders, which makes intensive communication essential. There is thus a question as to what level of knowledge about these processes should be expected from citizens wishing to participate. There are obvious differences between lay people and experts, again with a wide range of intermediate degrees of expertise.

Customer-oriented services require firstly that the users' expectations be known to the planners; secondly that they are given the necessary attention in

11 See. e.g., Hilgers (2003), Langweg et al. (2005).

12 Cf. LTUC (2002), for example, pp. 26, 35, 64.

13 'Not in my backyard'.

14 Weiß (2003).

the planning procedure. In this context, user participation aims to counteract typical patterns of behaviour or thought which can be found among product developers:

- first, *the lack of knowledge* about or *misinterpretation* of user needs;
- second, deliberate *neglect or devaluation* of these needs in the course of competing objectives.

Methodologies for user involvement for addressing the former concern can be divided into four approaches:

1. surveying of users' (and non-users') views using, for example, stated preference surveys on existing features or future changes of the service;
2. retrospective analysis of individual comments from customers; however, this does not provide a representative overview of their priorities;
3. discussion and assessment of services through participative practices (see Section 4.5);
4. experimental services, observation techniques.

In choosing from these instruments, one has to bear in mind their different suitability for various tasks and questions. Key parameters in this context are the complexity and number of details involved in the question and the relative familiarity of the topic or product.

Disregard or neglect of user needs can have various reasons. Some can be summed up as political factors in a wider sense (see previous sections). Passengers' needs are also not homogeneous, and reasons for different expectations can be based on factual requirements as well as subjective perceptions. It is no rarity to find different and even incompatible expectations in different groups.

To deal with such differences is a typical, recurrent planning task. Planning may even be defined as the attempt to reconcile conflicting demands for and effects of measures based on common, agreed objectives and to then identify the best overall solution. This implies the ability to recognize isolated interests and value them accordingly based on an understanding of the situation as a whole.

A wide range of tools for analysis and appraisal exists for this purpose, for example traffic flow models or cost-benefit analyses. Their setup as neutral, standardized practices often produces abstract results, which are not easily understood by unskilled people. Furthermore, their uncritical use can distract from the particularities of the problem under review, leading to solutions that are less suitable than they could be.

Good planning should always pay attention to the concrete situation; however, in everyday practice this is not always guaranteed. Sufficiently detailed information may be missing in the database, time not sufficient for a thorough analysis, or greater credibility given to the 'objective' result than voices from the public. In this situation, participative procedures can help planning to correct such errors and obtain additional information (see Section 4.4 and the example in Section 5.6.5).

1.2.3 Production: Considering Quality from the Customer's Perspective

Quality is defined as the degree to which a unit fulfils the demands of its inherent criteria (see Section 3.2). In public transport, the inherent criteria are mainly defined by the timetable on offer. In contrast to other utilities such as energy or water, public transport is not available continuously and ubiquitously. Hence users need to reconcile their travel needs with the service on offer which is defined mainly by the timetable, accessibility of infrastructure, and vehicles and other regulations or conditions. Consequently, passengers request services to operate as planned, published and advertised in order to be able to plan their movements reliably. This should be seen as a general principle, even if small diversions from the timetable may in some cases be considered as acceptable.

Key tasks of quality management are:

- services have to be *produced* in such a way that the targeted level of quality is reached in the most efficient way. This is done by means of internal quality management procedures which are in turn linked to norms and standards;
- the targeted level of quality has to be *defined and described* in such a way that the degree to which it is achieved can be measured. This is particularly important if bonus-malus payments or other consequences are linked with passing or failing these levels, as is often the case of transport service contracts (Section 3.5.1);
- quality standards should be *communicated*, because they inform (potential) customers about the level of service they can expect. This is especially important in services like transport, because services deemed unsatisfactory cannot be returned and not always be rectified through repetition. Possibilities to communicate these standards include certification of the provider and the marketing of service elements as 'guaranteed' (Sections 2.4 and 3.5.1);
- quality standards have to be monitored; the delivered level of quality and reasons for shortcomings have to be made public in order to enable the responsible bodies to take appropriate countermeasures.

Possibilities for users to influence these factors at the time the service is provided are quite naturally limited. Customers can, however, become involved in monitoring and maintaining quality standards, for example (for details see Section 3.5):

- in *consultations* about the standards and criteria in use;
- via customer satisfaction *surveys;*
- in user *panels* interviewed regularly;
- through better use of passengers' *observations* about acute problems.

Quality management is also linked with the other levels of user interests. First, reliable services require adequate operating conditions (a political and planning issue) and an internal structure on the operator's part which provides for a high-quality product. Second, the level of information, its detail and provision, directly affect passengers' possibilities to plan their trips and hence the level of precision they expect.

1.2.4 Practice: Taking Customers Seriously When They Have Problems

Despite the efforts taken in the process of quality management, public transport operates in a complex environment where various factors both inside and outside the operator's control can lead to service disruptions, missed connections and other problems. On the *practical level*, customers therefore request operators to secure suitable alternatives in case of malfunction or provide adequate redress and handle complaints in an uncomplicated way.[15]

From the users' perspective, a service malfunction must be understood in a wider sense as any deviation from those service parameters which had been advertised when the decision to travel was taken. These parameters are mainly advertised or determined in the timetable; however, not only delays but also early departure from stops and the lack of advertised features such as catering, first-class or disabled access should be included in this definition. In the same way, problems caused by wrong information given by the service providers should be covered, as they may cause problems very much like those from 'real' deficits.

In contrast with most other sectors, public transport operators are not required to compensate consumers for delays, cancellations and missed connections – the most common types of service malfunction. The factual reason for exempting the operators from the legal principles applied elsewhere was the extent to which rail and bus operations were exposed to risk of disruption caused by outsiders. Providers should not be held responsible for damages caused by others such as fellow road users or by natural disasters.

Against this absence of formal obligations is the commercial interest of operators to keep their customers and maintain their willingness to use the

---

15 CRUCC et al. (1996), LRPC (2000).

service again. Operators thus have to decide if they stick to their legal duties or offer redress or compensation on an ex gratia basis.

A more detailed discussion of the legal arrangements is provided in Section 2.6 and shall not be anticipated here. But it seems appropriate to take a look at the driving forces behind the sometimes heated debate that has been developing around this subject for some years now.

In principle it seems understandable when companies object to being held responsible for events over which they have no control. Depending on the definition of delay and in particular whether or not responsibility for consequential loss is included, compensation might become a significant expense. To some extent, this basic rule perhaps would also be accepted from the users' side. What probably causes resentment is the fact that reference to *force majeure*[16] by an operator *cannot be verified* by its customers and/or the fact that the legal framework allows providers to decline any responsibility for irregularities *irrespective of their cause*. This easily can be seen as unfair treatment and abuse of the company's position. If one also considers the problems and damages that can result from a service disruption, for example if flights or important appointments are missed, the frustration of the customers concerned is probably understandable as well.[17]

Furthermore, from the customer's perspective, the reason for a disruption is of little importance: Due to the characteristics of the service, public transport users tailor their travel needs to the service on offer – the timetable – and expect the provider to adhere to it in the same way they do. Any deviation forces them to react and change their arrangements, no matter what the cause.[18]

Ex gratia compensation may satisfy complainants individually. However, their main weakness is the lacking transparency for customers as to if and how much compensation they might receive and under which circumstances. Furthermore, if redress is only provided upon detailed request, it rewards those who are more courageous and persistent in asking. And finally, as discussed above, the institutional context of public transport provision (Sections 1.1.1 – 1.1.2) also provides comparatively little incentive to take precautions against irregularities or provide good customer care and complaints handling.[19]

Recent policy initiatives by the EU Commission have started to change this situation – most notably the Regulations on air passenger rights (no. 261/04) and rail passenger rights and obligations (no. 1371/07).[20] Some Member States

16 Technical term for events caused by circumstances beyond one's own control (also known as Acts of God).

17 See, for example, Callies (1959/1991), Godwin and Patterson (1999) and Sections 2.2–2.3 for the psychological aspects of such experiences.

18 Cf. LTUC (2002), pp. 73–4 Pro Bahn (2003).

19 Pro Bahn (2004, 2005), Landwell and Associates et al. (2003), p. 17.

20 Commission Communication 2005/46 sums up the different activities (cf. European Commission 2005).

have launched similar activities on the national level, so that the situation of transport users being 'without rights' might have disappeared in some years' time. But until then, many questions remain to be solved.

Chapter 2
Consumer Protection in Public Transport

2.1 Passengers' Rights and More
Martin Schiefelbusch

In recent years, public discussion has accorded far more attention to the sucject of passengers' rights than all other aspects of the customer-provider relationship. In the outline of this relationship on four levels presented in Section 1.2, politics, planning and quality management (facilities) form the basis of customer service and legal issues. However, there is reason to consider the question from the opposite perspective as well: at present, consumer protection is the topic to consider before consumer participation is introduced as a fully-developed concept.

This chapter does not deal only with legal issues: the main topic of the following sections is dealing with customers who would like to express concerns and criticism of a transport service. From the service provider's perspective, the next consideration is dealing with customer feedback (Section 2.2/2.3). Many questions should be resolved on this level; if this does not work, alternative dispute settlement with the intervention of an arbiter or similar person (Section 2.5) can help to avoid a complex legal dispute and prolonged frustration. For concrete problems, alongside legal regulations (Sections 2.6/2.7), voluntary consumer protection instruments should be considered. Particularly so-called service guarantees (Section 2.4) have been applied to an increasing degree in recent years.

2.2 The Role and Concept of Customer Feedback
Uwe Böhme

2.2.1 Passenger Complaints as an Opportunity

Changes in general circumstances, such as decreasing numbers of school pupils, increasing recreation traffic and reductions in government subsidies, are forcing many transport companies to tighten their belts. This need not always lead to a deterioration in service, but in most cases passengers complain of dwindling numbers of ticket counters and train conductors. Even worse, customer surveys have consistently ranked local PT along with DB AG's long-distance transport in the lowest places on the popularity scale.[1] As the results of the last Eurobarometer study showed, Germany is ranked third to last with respect to customer satisfaction on the international scale, finishing ahead of Italy and The Netherlands.[2] Whether it is possible to increase the number of passengers under circumstances such as these is doubtful.

Usually, cost reductions first affect the areas that are not absolutely necessary to maintaining operation. In complaint management, text modules are used to process customer complaints quickly and with as few resources as possible. But here it can be observed that the written responses to complaints are not always in line with the content of the original complaint. In a study conducted in 2005, the arbitration board Schlichtungsstelle Mobilität found out that customer letters to DB AG were not even answered in 13 per cent of cases, 8 per cent of responses did not address the complaint, and 17 per cent dealt with the matter only partially. Only about one in five responses was designated 'okay' (21 per cent).[3]

This gives rise to the question: what positive effects could be expected for the transport company if efforts in complaint management were intensified and increased attention was paid to addressing customer complaints? This question is completely justified if we assume that, on average, less than 1 per cent of passengers submits a complaint annually. This number may seem small but, considering that the number of passengers annually is often a figure in the tens of millions, annual complaints sent to the company can easily be a six-figure number.

The following explanations should help clarify that, in many situations, passenger complaints can be used as an opportunity. They are a relatively low-cost way to identify potential current deficiencies that would otherwise need to be determined through costly market research. Also, a customer-oriented response to a complaint could have a positive effect on the customer's overall satisfaction and on the company's image. After a short introduction about the origin of (dis)satisfaction, the effects of satisfactory responses to complaints will be discussed in detail.

2.2.2 Customer Satisfaction in the Focus of Public Transport

The concept of customer satisfaction has its roots in the market research of the 1970s.[4] In the transport sector, this factor had previously only played a small or nonexistent role. Quality was ensured rather through 'hard factors', that is,

1 See e.g. Blümel (2004), p. 7; EORG (2002), pp. 17 et seq.
2 EORG (2002), pp. 17 et seq.
3 Schlichtungsstelle Mobilität (2005).
4 See Borth et al. (2004), p. 11.

service was designed to meet basic needs.[5] It was assumed that the service need merely be supplied and that demand would more or less automatically follow.[6] The fact that today, even in the public transport sector, increased attention is paid to 'soft factors' like customer satisfaction is evidence that the basic circumstances have changed drastically.

However, the results of market research are only partially applicable to issues of public transport, since public transport is a special form of service. Whereas customers usually have an array of choices in many other service areas, their choice in PT is quite limited (cf. Section 1.1.1). Furthermore, frequent users of public transport are trainees, retirees and students, as well as the socially weaker population groups (referred to as 'captives'). For objective or subjective reasons, these user groups depend on public transport. Whether or not they lodge complaints, they will likely continue to use the service in the future.

This should not, however, lead to the conclusion that intensifying customer service is unnecessary just because enough people will always be forced into using bus or train services anyway. In fact, around 26 per cent of (regular and occasional PT) passengers in Germany own a personal vehicle[7] that they could use as an alternative if sufficiently dissatisfied. Other empirical studies show that for 94 per cent of train travel, an alternative form of transport would be possible.[8] It can be assumed that only satisfied customers will on the whole remain loyal to public transport.

But when are customers satisfied and what requirements must be met? What means are necessary to avoid customer dissatisfaction? These and other questions will be discussed in the following section, though dealing with customer dissatisfaction is the focus. Chapter 3 deals with customer satisfaction as a goal of quality management.

2.2.2.1 The origin of (dis)satisfaction The most common explanation for the origin of dissatisfaction is the C/D paradigm (confirmation/disconfirmation). It is assumed that customer dissatisfaction results when there is a perceptible discrepancy between expected service and actual service. Figure 2.1 explains this relationship.

The expectations result from past customer experiences with the same or similar services. Further influences on expectations are price, image, knowledge of alternative means of transport and oral communication. This received information stems from the customer's social environment and media communication on the part of the company. The service is determined by

5 See the service pyramid of user expectations (Schnippe (2000), p. 22) and the explanations of the Kano model in following section.

6 See Schnippe (2000), p. 205.

7 BMVBW (2004), p. 47.

8 See Siefke (1998), p. 52.

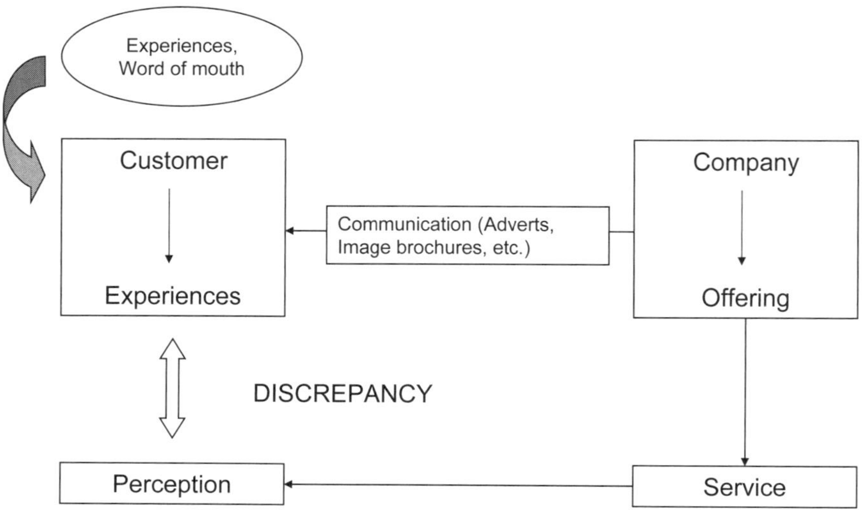

Figure 2.1 Origin of customer dissatisfaction

Source: author's depiction.

the company's offering and perceived by the customer during usage.[9] The customer conducts a target-performance comparison between the expected service and its confirmation or disconfirmation with respect to actual service,[10] which resulted in dissatisfaction in the last case.

The relationship between perception, expectation, and satisfaction is explained with somewhat more differentiation by service quality models. A frequently used explanation model is that of the tolerance zone, shown in Figure 2.2 below. Here it is assumed that customers develop expectations of barely acceptable and desired levels of service. Between these lie the tolerance zones.

If the service experienced by the customer lies within the zone of indifference, high satisfaction is certainly not the case, but rather minimal satisfaction or indifference, which manifests itself as a neutral emotional state. Only when expectations are exceeded does the customer feel excitement and high satisfaction. If the experience of service lies below the level that the customer deems barely acceptable, dissatisfaction results and can lead to complaints.[11]

---

9 Perceived quality also serves to measure satisfaction and is used to evaluate the quality of PT; in evaluation, measurement of the company's performance also has an influence (see 'Quality Loop' in Section 3.3).

10 See Siefke (1998), p. 65.

11 For more on the manner of customer complaints see Sec. 2.2.2.2.

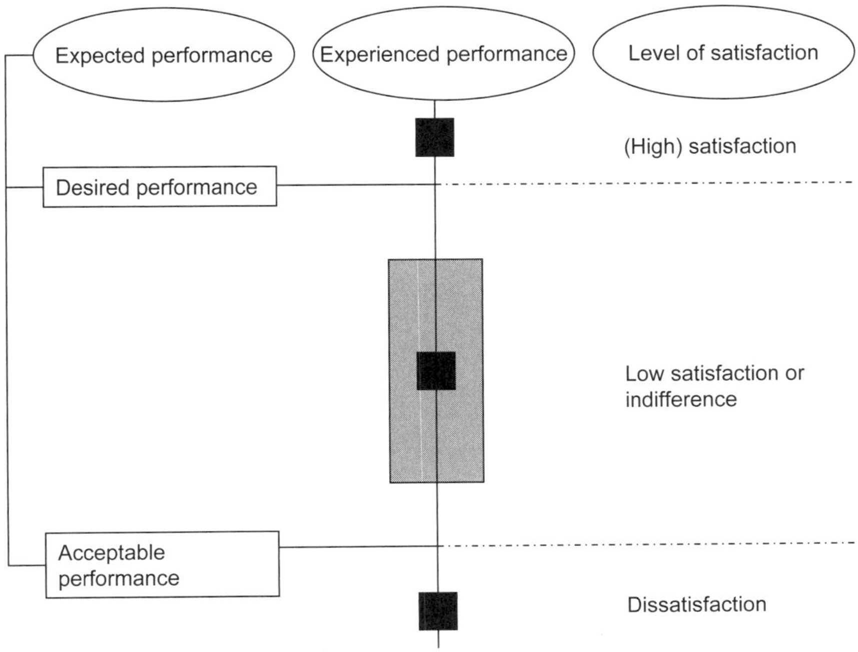

Figure 2.2 Tolerance zone model

Source: Stauss and Seidel (2002), p. 57.

This model provides an important reason for complaints: customers do not complain because the service they experience fails to comply with their ideals; they complain when their *minimum expectations are not met*.[12]

In this situation, the Kano model, shown below in Figure 2.3 can be helpful. Here the relationships previously described are represented graphically. Customer demands are divided into *basic*, *performance* and *enthusiasm* requirements. The x-coordinates symbolize the degree to which expectations were fulfilled while the y-coordinates show the level of satisfaction.

The customer presumes *basic requirements* like punctuality (within a range of tolerance), direct connections, as well as ease of interchanges. The lower curve shows that *even when basic requirements are met*, customer satisfaction *still lies in the negative area*, only peripherally cutting into the zone of indifference. It also shows that *considerable dissatisfaction occurs as quality falls*.

The upper curve shows that *enthusiasm requirements*, which were not explicitly expected, even with a low degree of fulfilment, have a clearly positive effect on satisfaction.

12 Stauss and Seidel (2001), pp. 57 et seq.

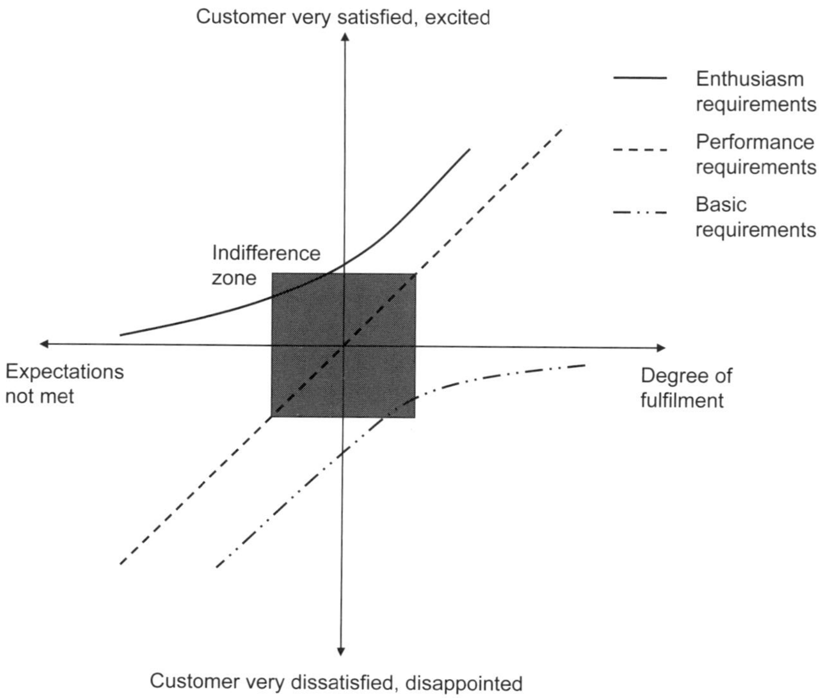

Figure 2.3 The Kano model of customer satisfaction

Source: Bracher et al. (2004), p. 133.

The service requirements in PT are today understood to include the quality of the experience of waiting at stations, friendliness of personnel, and purchasing tickets at ticket machines.[13] If these are offered by many companies, they could also become basic requirements. On the other hand, one should not delay the introduction of enthusiasm requirements that the customer does not (yet) expect, since customer expectations are raised by such introductions. Newspapers or television in local public transport are considered enthusiasm requirements. However, the introduction of additional services such as these is only sensible once all other requirements have been fulfilled.[14]

With regard to this point, however, it should be mentioned that the degree of satisfaction represents a necessary, but in no way sufficient criterion for the use of public transport. In this context, Reichheld[15] speaks of a 'satisfaction trap' in which, despite a high degree of satisfaction, customers drift away due to marginally better competitor products. One could speak of true customer

13 See Topp (2004), p. 475.

14 See Topp (2004), p. 475; Probst et al. (2001).

15 Reichheld (1997), pp. 61 et seq., quoted in Schnippe (2000), p. 149.

loyalty if continued purchases (continued use) were voluntary and not due to a lack of alternatives.

One possibility for maintaining customer loyalty that has seldom been exploited up to now is to pay more attention to customer complaints. Knowing the basic principles mentioned and an efficient method to bring implement them in complaint management can contribute to improved customer retention. First, one must deal with the *manner* of the complaint before going further into the content and goals of complaint *management*.

2.2.2.2 Manner of customer complaints The relationship between customer loyalty and experience reports was first studied in the 1970s. TARP surveys[16] funded by the American government analysed the manner of customer complaints in representative polls and refuted the commonly held position (especially in the business world) that a small number of complaints indicates a high number of satisfied customers. For example, a central finding of the study was the following:

> Only about four percent of unsatisfied customers file a complaint with the company. For every complaint received by the company, there are 26 customers with the same problem who are not speaking out.[17]

Other results indicate 'for every openly expressed complaint, there are six severe complaints and 20–50 less severe unvoiced complaints'.[18] This relatively high number of unvoiced complaints can be explained by the various aspects that keep customers from submitting complaints.

In the first place, customers are unsure if their subjective perception of poor service warrants a complaint. A contributing factor is also the lack of hard proof, since the service is intangible and the simultaneity of supply and demand of the service (*uno actu* principle) is not storable.[19] These initial circumstances, taken together with the comparatively high complexity of a complaint with no guaranteed success,[20] often result in customers reporting their negative experience only to friends and acquaintances rather than to the company (see below).

Another reason unsatisfied customers do not file complaints could have to do with artificially installed corporate barriers. Examples of this might be service hotlines that require payment or addresses that are difficult to find on the Internet. Such obstacles can cause a company to miss out on chances

16 Technical Assistance Research Program, see e.g. Murphy (1994), pp. 8 et seq.
17 See Schnippe (2000), p. 156.
18 Haeske (2001), p. 17.
19 See Schnippe (2000), p. 156.
20 As long as there is no legal claim for financial reimbursement (compensation), the customer can only rely on the company's goodwill.

to hear customers' opinions. While the number of unsatisfied customers should be minimized, the number of unsatisfied customers who actually voice their complaints should be maximized.[21] Communication channels have to be more easily accessible in order to encourage customers to articulate their complaints.

The manner of customer complaints can appear to be quite complex. In addition to the complaint itself, other forms of exposure to dissatisfaction are to be found: by word of mouth as described below, but also by avoidance or 'acts of revenge'.[22] Researches disagree as to whether dissatisfaction can be regarded as a necessary requirement for complaint. On the one hand, dissatisfaction can result in no reaction whatsoever, but on the other, some customers articulate complaints despite being satisfied.[23]

One particular facet of the manner of complaints that has an especially high potential of improving the company's image (but also a high risk of worsening it) is articulated in the second central finding of the TARP study:

> A satisfied customer shares his/her experience with around three people, while an unsatisfied customer tells around ten people about his negative experiences. In addition, customers whose complaints[24] were resolved to their satisfaction tell an average of five people how the company handled the situation.

This result makes clear two essential characteristics that should be observed when dealing with customer complaints. First, word-of-mouth should assume a more important role, since it is considered more effective than mass media advertising.[25] Second, it displays the initially contradictory-seeming phenomenon of the complaint satisfaction paradox (see Figure 2.4). According to this principle, overall satisfaction is decreased when a negative event occurs. If the customer does not complain, his/her degree of satisfaction stays roughly at this low level. If the customer contacts the company and is satisfied by the company's response, then this complaint satisfaction can be greater than the overall satisfaction before the occurrence of the problem. The end result is that the loyalty of customers who have received favourable

---

21 See Stauss and Seidel (2002), p. 51.

22 Godwin and Patterson (1999), Huefner and Hunt (1994).

23 See Borth et al. (2004), pp. 31 et seq.

24 A claim is a special form of customer complaint that deals with objections to products or services that have explicit or implicit purchase agreement demands, which might be upheld in a court of law. In public transport, the foundation for such demands is lacking in many cases (see Sections 2.4–2.7). Thus in most cases related to public transport, it is a matter of complaints and not claims. It is assumed that the range described also applies to complaints.

25 See Borth et al. (2004), pp. 156 et seq.

responses to their complaints can be greater than that of those who have never had a problem with a product or the company.

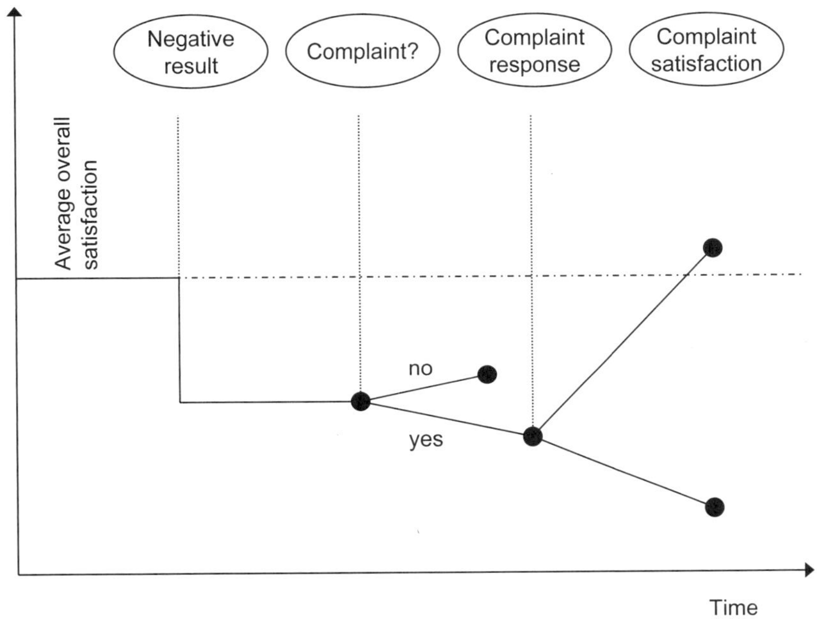

Figure 2.4 Effect of complaint satisfaction on overall satisfaction

Source: Stauss and Seidel (2002), p. 75.

Though this phenomenon has been proven in several studies,[26] it should not be regarded as a universal truth.

First and foremost is the question: to what level can customer satisfaction fall and, with an appropriate response, still be raised above an average level of overall satisfaction? Because even the best complaint management process cannot make up for a very poor product or an unsatisfactory service.[27]

Conversely, a very poor complaint response can result in a double deviation that strongly damages the customer relationship, thus drastically reducing the chances of customer retention. Given that it can cost as much as five times as much to win a new customer than to retain one,[28] complaint management is an outstanding and comparatively inexpensive means of increasing customer retention.

26 E.g. Björlin-Lidén (2004)
27 See Borth et al. (2004), p. 47
28 See Haeske (2001), p. 21

The next TARP study finding also speaks to the value of an effective complaint response:

> Customers who have filed a complaint – regardless of the outcome – are more likely to purchase services or products from the company again than those who have not filed a complaint.[29]

This statement also makes clear the motivation behind a complaint. When a customer makes a complaint, this is partially an expression of their interest in the company, as they are intentionally making a suggestion for improvement.[30] Often, however, companies treat customers who have filed complaints as opponents, categorizing them as 'whiners' or 'belly-achers' who, by filing partially illegitimate complaints, result in high costs.[31] In most cases, however, this is a false impression, as customers who file complaints are often the most loyal customers[32] who deserve special attention in complaint management.

2.2.3 Goals and Content of Complaint Management

A complaint is the realization of an unmet expectation[33] that is expressed in order to draw attention to a behaviour or action of the service provider that elicited a subjectively negative response in the hope of securing compensation for damages suffered and/or effecting future changes.[34] The primary goals of complaint management are to increase profits and competitiveness, to restore customer satisfaction, to minimize negative effects of customer dissatisfaction on the company, as well as to make use of the information contained in complaints[35] as regards deficiencies and drawbacks.[36] Complaint management (CM) contains several steps in effectively addressing these goals. These include complaint stimulation, receipt, processing, response and analysis.[37]

---

29 Schnippe (2000), p. 156.

30 See Stauss and Seidel (2002), p. 52.

31 See Stauss and Seidel (2002), p. 52.

32 See Stauss and Seidel (2002), p. 52.

33 See Barlow and Møller (1996), p. 23.

34 See Borth et al. (2004), p. 29.

35 Complaints often contain large quantities of critical remarks. However, complaints could also contain praise, if even sparsely. For this reason, the German Road and Transportation Research Association (FGSV) suggests using the term 'praise and complaint management' in order to allow positive aspects to influence the evaluation of the quality of service. (see FGSV (2004), p. 125).

36 See Stauss and Seidel (2002), p. 79.

37 See Meffert and Bruhn (2003), pp. 415 et seq.

While the first four steps constitute direct CM, complaint analysis is regarded as indirect CM,[38] since it deals with internal business processes that aim to identify consequences from customer complaints, in order to better adapt services to customers' needs.

Complaint stimulation in *direct* CM is designed to encourage dissatisfied customers to voice to the company the problems they experience. Installing easily accessible (medial) complaint channels is as important as their communication. For example, free service hotlines encourage additional complaints while also increasing consumer trust.[39] In addition to the possibility of sending complaints by email, the postal address of the company should be given on the website. For many different reasons, some customers who have Internet access might still prefer to send complaints by mail. For example, they might be dissatisfied with the result of an oral complaint or simply feel more comfortable sending a written complaint. Even if customers want to take legal action on account of an especially severe grievance, a written complaint is a possibility.[40]

The *complaint receipt* phase deals with the organization of complaint entry. In the case of a written complaint, initial contact with the customer in the form of a receipt confirmation should occur during this phase. Especially when the anticipated processing time is long, an intermediate reply is recommended, so that the customer is informed about the whereabouts of their letter.

Complaint registration should record the problems mentioned by the customer quickly, completely, and in a structured manner. Complaints sent electronically can be organized in a standard form letter and/or with fixed-entry blanks.

The *complaint processing* phase deals with determining accountability, defining processing deadlines, or installing mechanisms for monitoring deadlines.

At the end of direct complaint management comes the *complaint response*, in which the customer receives a letter of reply articulating solutions to the problem and compensations or apologies. Except in the case of an intermediate reply, this represents the first customer contact[41] when dealing with written complaints. Given the desire to generate a high level of complaint satisfaction with the customer, complaint responses are of tremendous importance.

---

38 Stauss and Seidel (2002), p. 82 subdivide indirect complaint management further into complaint analysis, management controlling, reporting and use of information.

39 See Barlow and Møller (1996), p. 197.

40 See Barlow and Møller (1996), p. 148.

41 The customer experiences the 'moment of truth' as soon as the complaint letter is sent off and the time of waiting for a reaction begins (Stauss and Seidel (2002), p. 84).

2.2.4 Conclusion: Complaint Management as Underestimated Resource?

The theoretical part of the explanation showed that, due to the larger impact of negative experiences, customer satisfaction has a decisive influence on a company's image and therefore also on a company's success. Since public transport is increasingly involved in competition, 'soft factors' like customer service are more important than they have previously been. Fewer and fewer people depend on public transport and will use bus and train services only if they result in high satisfaction. This cannot be attained, however, merely by fulfilling minimum requirements like punctuality or cleanliness, since such aspects are prerequisites for most customers. Much more effort is required to retain customers for long periods of time.

Complaint management can be employed as an inexpensive means of customer retention. The fact that customers whose complaint was resolved to their satisfaction exhibit higher overall satisfaction than customers who did not feel motivated to make a complaint exemplifies the benefits of customer-oriented complaint management.

Although complaint management along with other factors contributes to improving a company's image, sometimes even resulting in increased overall attraction, it still receives too little attention. Customers are dubbed 'whiners' or 'belly-achers' and some companies still erroneously conclude that a relatively small number of complaints indicates a small number of unsatisfied customers. This is, however, only partially correct: one the one hand, the number of unsatisfied customers must be minimized but, on the other, the number of unsatisfied customers who actually lodge complaints must be increased. The difficulty lies in processing complaints quickly while creating individual complaint responses with appropriate compensatory measures – in spite of high costs – in order to please dissatisfied customers. This is no easy task and, as is shown in the following section, the cost factor is often the focus of complaint processing, which unfortunately obscures the goal of generating customer complaint satisfaction.

2.3 Complaint Management in Practice – Models of Argumentation and Customer Expectations for Complaints

Uwe Böhme

2.3.1 Background

As many cases demonstrate, complaint management often still has shortfalls in terms of customer-orientation. For instance, a study carried out by the Schlichtungsstelle Nahverkehr NRW (local transport arbitration board, cf.

Section 2.5)[42] on the complaints handling of German railways showed a high potential for improvement regarding various criteria such as promptness, individuality, phrasing and customer-friendliness.

In the following section, a special focus is placed on complaints, because in complaints one can see customers' expectations for complaint response. This is based on the assumption that a comparison can be drawn between satisfaction in general and satisfaction with the handling of complaints as well as between expectations and their fulfilment (see Section 2.2). A strong orientation toward customer expectations in the written replies can therefore contribute to satisfaction with complaint management and in turn increase global satisfaction.

On this basis, complaints were examined according to expectations and finally evaluated in terms of the degree to which the respective replies met (answered) these expectations. Furthermore, common argumentation models will be shown through examples of correspondence between transport companies and passengers and recommendations will be given for the avoidance of mistakes in answering complaints.[43]

2.3.2 Results

2.3.2.1 Style and extent of text It can be observed that passengers' complaints are seldom written in response to a single instance, but that writing is the result of accumulated frustration created by several instances:

> I have often been annoyed, and also have often heard grievances from others, but one always hopes that in the end it will get better. For me, last weekend was the last straw and I am extremely furious.

Thus complaints come to consist of several A4 pages in which, alongside the actual complaint, earlier troubles are described which were not the subject of complaints. The comparatively large extent of the text is also due to the fact that, alongside pure facts, personal feelings, appeals to the company as well as personal attitudes are often expressed.

In the above example, it is also clear that the customer sees him- or herself as a representative of the customer group and places not only his or her own interests in the foreground. It is mentioned that other dissatisfied customers have expressed grievances as well, but have not complained to the company. It also becomes apparent in the use of the 'we' form that the problem lies not only with one person, but that a number of passengers exist with the same problem.

42 See Verbraucherzentrale NRW (2003).
43 See also LRPC (2000).

Written replies sometimes comprise only a few sentences that consist primarily of impersonal standardized phrases. However, in the second attempt, if the customer receives another written reply to a follow-up complaint, the answer is often longer.

There is some evidence of an effort to produce a balance between standardized text and freely composed text. Especially the replies to follow-up complaints tend to contain fewer stereotypical patterns than initial answers to complaints. At the same time, these replies often employ a complicated and bureaucratic writing style. Thus, along with very long sentences, technical expressions are often used that hardly contribute to understanding. Examples for this are 'the trains' operational performance' as well as 'factors influencing train movement'.

2.3.2.2 Reference to customer requests Complaints are for the most part addressed generally ('To Whom It May Concern'). Some cynical phrasings are also used from time to time such as 'To The Highest in Command', and occasionally the head of the company is personally addressed. This can be taken as an indication that customers are not aware who they should address in the case of a complaint. There is also an express desire in these incidences to personally speak with the head of the company.

While addressing indirect issues is not explicitly called for, it can often be observed in *follow-up complaints* that hidden circumstances should have been considered. Thus for example a customer complained that due to deficient intermodal coordination and lack of timely information, a transfer to a connecting bus was not possible. This complaint includes first a question regarding the possibility of compensation, the consequences for the employees who are to blame, and a request for a review of the incident.

Included in the written reply was a travel coupon good for €10. The reply to the question was, however, limited to a notice that in cases of delay, it is not possible to issue cash compensation. The subsequent complaint shows that it was not only expected that the direct questions would be answered, but that reference should be made to further points:

> (…) your hope, however, deceives you: I am not a satisfied customer. You did not (despite the long processing time!) address my letter at all. Did you review the incident described? Where, then, were the weak points? Were the employees involved informed in order to prevent something like this from happening again? Do they know at all what they have perpetrated? Have measures been taken to inform travellers of cancellations in the future? How can the cooperation between train and bus travel be improved in such critical situations? You could have given an explanation of such issues (you did indeed give yourself a good amount of time).

In contrast to the complaint, where the customer can often not choose a personal address, the *written reply* often employs the following salutation: 'Dear Ms (Mr) ..., Your trip from ... to ... did not proceed as you might have expected.' If the complainant wrote to the head of the company, the first sentence next makes reference to the form of address. This explains to the customer that the head of the company requested that this associate handle the customer dialogue.[44]

Even if the complainant's questions are addressed, it is sometimes difficult to find a reference to them in the response to the complaint. Especially in cases where several points are addressed, references to the passenger's concerns are only seldom made clear in the introductory sentences. The requests are often not explicitly repeated, but referred to generally as the 'resulting inconveniences'.

2.3.2.3 The goal of complaints One of the goals of complaints is to inform the company of deficiencies and to find out about future improvements. In some cases this could be identified as the main concern, making financial compensation a low priority. Especially in complaints that have addressed several points and, for example, have received a coupon for a low value in response, this perspective can be observed in the follow-up complaint. However, one study[45] determined that financial compensation has a positive influence on satisfaction in complaints.

People often try through various verbal means to apply pressure on the company to provide compensation and/or compensation. This can come in the form of threats to switch to other means of transport or of references to the fact that one has a customer card and thus is a frequent traveller and would like to be treated accordingly.

Showing cooperation is also common. This sometimes expresses, for example, that further claims (although fully justified in the customer's view) will be foregone if the company shows according goodwill.

On the company's side there is a definite readiness to include a coupon with the written reply in individual cases, even if there is no compensation payment stated in the terms of carriage. It cannot, however, be said how much influence the aforementioned threats have.

The response to passengers' concerns about existing deficiencies often includes apologies and expressions of regret, as well as insight into the problem. The latter make clear, for example, that 'we have not yet reached our goal and have to invest more in training our employees in the future'. In this context, there is often a pattern of argumentation that seeks to put the company's qualities in focus. Thus it is stated, for example, that travel disturbances cannot

---

44 Especially if larger companies are involved, it can be doubtful that customers will believe this statement.

45 See Baer and Hill (1994), p. 148.

be avoided but, 'it is notable that 90 per cent of our trains are, however, on time'. Their regret is also expressed when a claim for compensation is refused. A request for the customer to continue travelling with the company in the future is often included.

2.3.3 Recommedations for Practice

Examination of customer expectations for written replies evinced three central aspects that are to be considered customer-oriented complaint response.

1. *Longer written replies are seen as a sign of interest.* Research shows that recipients of longer written replies react more positively than recipients of shorter ones. They see these as a sign of interest.[46] The usage of standardized text could be a sensible addition to the efficient processing of customer complaints. However, addressing the individual as well as providing the fastest possible answer should always be in the foreground. It is, however, also to be observed that in formulating a complaint response, the customer's satisfaction with the complaint process is decisively influenced not only by content, but also by linguistic style, as well as grammar and orthography. Mistakes of this kind lead to the customer seeing the handling of their complaint as careless. This can often lead to negative conclusions being made about the company's complaint management, and trust being totally lost. Written correspondence has the particularity that when misunderstandings arise, there is no direct inquiry possible, this can only be accomplished with a subsequent letter. For this reason, complicated sentence structure and the use of technical terms are to be avoided.
2. *Indirect issues must also be addressed.* In every case, all aspects of the complaint must be addressed. It is often shown in practice that, for instance in claims for financial compensation, they are often granted without reference to directly or indirectly posed questions (reasons for delay). In cases where several points are expressed, it should be recognizable which concern is being addressed. The introduction should begin with an explicit restatement of the problem. This communicates to the customer that the problem is understood in its full extent.
3. *Financial compensation is not the main concern.* Since the DB customer charter was enacted in 2004 and specific regulations for the amount of compensation have existed, it has been possible to defer priority in complaints about delays to a claim for compensation. While the inclusion of coupons in written replies have a fundamentally positive influence on complaint satisfaction, financial compensation is not sufficient to produce satisfaction (see 2. above). Many complainants

46 See Barlow and Møller (1996), pp. 158f.

are more interested in an adequate apology, a good explanation and measures toward a solution.[47]

In denying requested claims for financial compensation, it is recommendable to avoid formulations such as '… we would be glad to have the opportunity to welcome you as customer in the future' because in cases of refusal these can be seen as scorn and lead to frustration, in general producing a feeling of perturbation. It is much more advisable to once more express regrets about the incident, for example to ask for understanding for the decision.

2.4 Service Guarantees and Customer Satisfaction – Application and Experience in Germany

Nancy Neugebauer

Early deregulatory reorganization of the market and the resulting competition make it necessary for transport companies to rethink how they can provide their services with a stronger passenger orientation. It will be increasingly necessary for transport service providers to address customer needs and to monitor delivered service quality to be able to compete.[48] At the same time, claims for improved customer protection in cases of service disruptions are increasing in the field of consumer politics. The customer's current legal position in cases of delays or cancellations (see Section 2.6) is being increasingly criticized as insufficient.

This is the basis for the following discussion of service guaranties in theory and practice.

2.4.1 Evolution of Service Guarantees in Germany

In 1997, the Berliner Verkehrsbetriebe (BVG) introduced the first customer guarantee in Germany. Since that time, the number of transport providers offering guarantees has been increasing. In Germany today,[49] seven transport associations and their member companies, as well as 44 transport companies, guarantee their customer-related service standards 'on their own initiative'.[50] These companies are not evenly dispersed through the country.[51] Hesse, North-

47 See LRPC and CRUCC (1996), p. 20.

48 See Czech et al. (2002), p. 25.

49 As of January 2007.

50 The empirical data used in this article about service guarantees for public transport in Germany are based on a study conducted in the context of the diploma dissertation of the same name by the author; see Neugebauer (2006).

51 If one investigates the service areas for the respective transport companies that offer service guarantees, it can be seen which regions in each state already offer

Rhine Westphalia and Brandenburg contain the majority of districts and cities with transport companies that offer guarantees. In the city/states Berlin and Bremen, local transport companies guarantee their quality standards as well. The punctuality guarantees for all local passenger rail transport (LPRT) in Schleswig-Holstein should be highlighted as well. In Baden-Württemberg, Bavaria, Mecklenburg-Western Pomerania, Lower Saxony, Saxony, Saxony-Anhalt and Thuringia there are, in contrast, only a few guaranteed services, while in Hamburg, Rhineland-Palatinate and Saarland, there are to date no such offers. In reference to long-distance passenger rail transport, it should be noted that Deutsche Bahn AG's customer charter with its compensation regulations in cases of delay is valid for the whole country. Because DB AG's competitors only serve isolated routes, almost all long-distance train services within Germany are covered by a service guarantee in cases of delay.

In these times of increasing competition and growing demands, this situation raises the question of what potential can actually be accorded to service guarantees in terms of passenger protection in the case of delay. What effects are there to be seen in how companies and customers have used guarantees? It should be determined what should be addressed as service guarantees specifically in the field of public transport. What concepts and experiences are there to be found in Germany? The next step is indeed to define when one can speak of a service *guarantee*.

2.4.2 *What is a Service Guarantee?*

Two approaches to quality promises can be distinguished. First, with the help of *quality and/or performance promises* (also known as charters), the company can communicate the standards that it strives to fulfil. These performance criteria are meant to contribute to demonstrating to the customer that the company is highly dedicated. This includes, for example, ensuring friendliness, safety, and engaged complaint management. If these promises are not fulfilled, there are, however, no direct consequences for the company or the customer.

In contrast, *service guarantees*, also called customer guarantees, define quality standards in the service provision process for which customers can expect *compensation*; one has a claim to monetary or other forms of reparation. For example if punctuality is promised, the customer's fare is (partly) reimbursed in the case of delay.

Figure 2.5 demonstrates the conceptual division between a promise and a guarantee. By continuation, the service guarantee's regulations can be

---

LPT users a service guarantee. Transport companies' service areas are mostly based on district boundaries (regionalization law). It cannot be ruled out that several transport service providers are active in one district or city/state. There can be no conclusive statement as to whether all of the companies active in a region offer a service guarantee; see Neugebauer (2006), p. 45.

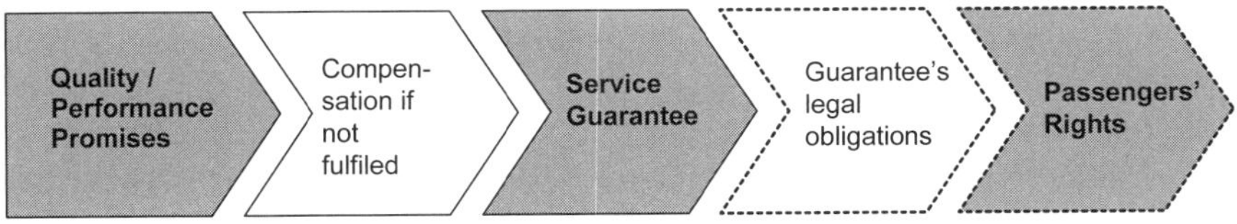

Figure 2.5 Delineation of the legally binding nature of quality promises

Source: author's depiction.

integrated into the transport company's respective terms of carriage. This would make the promised means of compensation legally binding in cases where the announced standards are not fulfilled.

In Germany, almost half of the transport providers compliment their punctuality and/or connection guarantees with various quality promises such as, 'we will inform you quickly and completely' (SSB) or, 'we strive to process your concerns in a friendly and timely manner' (SHA). However, only the guarantees from the transport association VRR and VRS and the compensation offered by DB AG in cases of delay are posed in their respective terms of carriage such that the customer has a legal claim. All other service guarantees are voluntary commitments by the transport provider and are thus not legally binding (see Section 2.6).

In what way then is the offer of a service guarantee useful? What advantages does it provide the customer? What effects can the company expect? And what role do service guarantees play in the transport authority-transport company relationship in public transport?

2.4.3 Effects of a Service Guarantee

The effects of a service guarantee can be observed from three perspectives. While it is primarily a sign of quality for the customer; for the company it presents a potential instrument of quality control. In the relationship between the service provider and the transport authorities, on the other hand, the guarantee can have a supervisory function. The various effects can be described as follows:

2.4.3.1 A sign of quality from the customer's perspective Customers have to deal with the problem of not being able to judge the quality of a service before using it; that is that passengers do not know before they board a vehicle whether or not they will reach their destination on time. The extent and quality of service cannot be foreseen because the service is delivered individually for each customer – production and consumption coincide. Appraisal of a service is thus for the most part dependent on experience and trust. Because of these factors, *it is necessary to show the customer the quality and reliability of a service before usage in order to reduce the perceived purchasing risk.* [52]

A service guarantee thus seems a suitable instrument because it communicates the definite quality standards that the customer can expect. It *decreases the user's uncertainty*. In addition, the perceived purchasing risk reduces the prospect of compensation in the case of poor performance from a provider. Users reason that a company only offers a service guarantee when it is actually sure that it can maintain the promised standards and are thus convinced that they will receive the service with the expected quality.[53]

Thus, by providing a sign of quality, the application of a service guarantee can potentially win former non-users, because they can now evaluate the service quality more easily before usage.[54]

For existing customers, a service guarantee increases satisfaction and brand loyalty.[55] The communication and implementation of quality standards as well as the service improvements caused by a guarantee confirm for the customers that they have chosen the right provider. Bockholt and Probst[56] show additionally that customers are prepared to pay more for guaranteed services. It was possible to prove that customers will accept an addition of 24 cents for the institution of a guarantee for night-time connections. The prospect of a positive outcome and a reduction of negative consequences increase the value of a service, which is shown by customers' *increased willingness to pay*.

What effects does a service guarantee have on the satisfaction of customers who have made a claim on the guarantee after receiving poor service? Customers are generally understanding when irregularities in service occur. However, when further problems occur in the course of compensation, dissatisfaction rises considerably. It can thus be reasoned that the compensation process takes on an important role because of the negative effects that a second problem pose. Björlin-Lidén[57] demonstrates in her research that *satisfaction* was *higher* for customers of Stockholm local transport after making a claim on a guarantee than *before a problem occurred*. This phenomenon of the recovery paradox (see also Section 2.3) occurs in situations where customers have had a problem that could be satisfactorily solved. After a successful compensation process, they rate their satisfaction as higher than if the problem had never occurred. This can be ascribed to the fact that in making a claim based on the guarantee, the customer receives the possibility to share their displeasure with the company (by telephone or by submitting a guarantee form). This provides

52 See Meyer and Blümelhuber (1999), pp. 203 et seq. and Bruhn (1996), pp. 11 et seq.

53 See Wirtz (1998), pp. 832 et seq.

54 See Bockholt and Probst (2003), p. 25 and Wirtz (1998), p. 833.

55 See Bockholt and Probst (2003), p. 25 and Wirtz (1998), p. 833.

56 See Bockholt and Probst (2003), pp. 26 et seq.

57 See Björlin-Lidén (2004), Paper V, pp. 3 et seq.

the *feeling that their opinion is important to the company*. In addition, receiving compensation increases the customer's satisfaction.

At the same time, the customer feels fairly treated if the receipt of reimbursement is a result of standards that are anchored in company politics. In the processing of customer complaints and the valuation of compensation on an individual basis, a disadvantage is the lack of *transparency for the criteria* that have led to the company's accommodation. The customer thus cannot judge whether he can count on the same goodwill from the company if he experiences another problem in using the service. Also, this offer only comes into effect upon direct demand.[58]

2.4.3.2 From the company's perspective, a quality control tool From the company's perspective, a service guarantee fulfils an essential role in service communication by giving signs of quality and providing active complaint management.

When customers complain, they express their dissatisfaction. Indeed, not every dissatisfaction leads to a complaint. Often, dissatisfaction remains as an unvoiced complaint which can be expressed by switching to a competitor or through negative communication to third parties (see Section 2.3). It is thus worth motivating customers to communicate their dissatisfaction; in *knowing the company's weak points* there is an opportunity to correct them and at the same time to satisfy the customer, even if retroactively.[59]

Before customers express their dissatisfaction to companies, they evaluate their complaint's prospects of success. This is often not a source of evidence for bad service. In addition, customers often apply their own subjective/individual criteria.[60] It is thus unclear, for example, if a delay of 10 minutes merits a complaint or if it takes a 30-minute delay. With a service guarantee, clear quality standards are signified and the customer can judge what prospects of success a complaint will have and what compensation they can expect.

When claims on the guarantee come in, the company receives *direct feedback about the service quality they offer* and thus has the opportunity to identify weak points in their system.[61] The customers thus become 'the permanent watchdog for service quality',[62] and complement complex bureaucratic quality safeguarding systems. According to the Managing Director of the German transport association[63] NVV, the association's motivation for instituting a five-minute guarantee is that 'for us, every response is an incentive and a

58 See Björlin-Lidén (2004), Paper V, pp. 10 et seq.; Schiefelbusch (2004), p. 320.

59 See Bruhn (1996), p. 7 and Wimmer and Roleff (1998), pp. 267 et seq.

60 See Wimmer and Roleff (1998), p. 275.

61 See Hart (1988), p. 58; Laumann et al. (2002), p. 293; Wirtz (1998), p. 831.

62 See Bockholt and Probst (2003), p. 25.

63 The term 'transport association' is used in the following to refer to a German Verkehrsverbund (see Sec. 5.8.2.2 for details) and similar bodies in other countries.

requirement to be even better. It is indeed our goal to detect and eliminate deficiencies which we might not have ever discovered by other means'.[64]

Alongside detecting deficiencies in operational activity, a service guarantee presents a further *impetus for increasing performance quality*. With the help of a guarantee, quality deficiencies are expressed monetarily. The costs created by providing the guarantee service itself, as well as the costs of guarantee processing can be calculated as part of the sphere of responsibility (cost centre).[65] There is thus an internal pressure on the individual departments to maintain guaranteed performance standards diligently and to orient the overall service provision system towards keeping the total operational expenses for the compensatory process as low as possible.

This necessitates an exact analysis of the weak points in order to avoid deficiencies permanently and achieve a sustainable *improvement in the service provision process*. At the same time, the companies are obliged to consistently concentrate on the customers' needs and expectations because if they do not keep in step with changing customer needs, it will lead to dissatisfaction, which is shown in increasing guarantee claims.[66]

The quality standards defined in the guarantee also show employees what service the company expects and creates a *guideline that personnel must follow*. If the employees, through training, are familiar with the defined standards and their potential effect on optimizing operation, they will become aware of their own importance in customer satisfaction. At the same time, a service guarantee makes *employees' performance internally visible* and increases employees' pride in their work.[67]

It can thus be ultimately determined that the goal of effective complaint management is initiating a customer dialogue in which customers are satisfied through receiving fast, individual and thorough replies to their concerns. The company, for its part, gains the possibility to learn through complaints and/ or claims on the guarantee how they can take the necessary steps to avoid the problems that caused complaints in the future.[68]

2.4.3.3 In the transport authority-transport company relationship, a quality control tool Service guarantees can have a further function in the relationship between transport authorities and providers of public transport services. In the transport contracts between public transport authorities and the respective transport companies, there are often quality standards stipulated, for example

64 See the NVV's press release from May 2006. http://www.nvv.de/Menus-rechts/Presse/Mai06_5Min.html (accessed February 2007)

65 See Baumeister et al (2005), p. 487.

66 See Wirtz (1998), pp. 829 et seq.

67 See Hart (1988), pp. 57 et seq.; Laumann et al. (2002), p. 293; Wirtz (1998), p. 832.

68 See Laumann et al. (2002), p. 293.

that 70 per cent of the stops must be equipped with SOS telephones or that buses must maintain 90 per cent punctuality, with a three-minute allowance. If these criteria are fulfilled, the company receives bonus payments; if the company fails to maintain the stipulated quality standards, they must make malus payments (see example in Section 3.5.1).

These stipulated quality criteria between the transport company and the primary customer (the public transport authority)[69] refer, however, to the company's (annual) overall performance and do not help the individual passenger in the concrete situation of a delay. In order to protect the end customer (the passenger), service guarantees need to be incorporated into the bonus/malus regulations to, for instance, require penalty payments if a certain number of guaranteed services is lacking. This would necessitate the contractual specifications to be converted into suitable performance guarantees. Claims on the guarantee would accordingly signal non-compliance with the contractual obligations. Guarantees could thus *replace cost-intensive contract auditing* as the fundamental element of transport contracts. With the evaluation of claims that are received, the transport authority, on the one hand gets timely feedback about the performance quality delivered by the transport company and, on the other, integrated and communicated service guarantees create *continual performance control* that are a permanent part of company policy as opposed to complex random-sampling-based customer satisfaction measurement.[70]

The survey of the transport providers that offer service guarantees in Germany, however, showed that to date only two of the 25 companies asked have bonus/malus systems that take the guarantees into account.

As Figure 2.6 shows, service guarantees can fulfil several functions. For companies, they present a tool for quality safeguarding in which customers check service quality with their feedback. For customers, the perceived purchasing risk sinks with signs of quality. The passenger receives a clear basis for judgement of performance so that when the service is unsatisfactory they know whether or not their complaint will have a successful result.[71]

If quality standards are set in a service guarantee, the questions arise: how does one distinguish an effective guarantee? What aspects must be provided to make a service guarantee attractive for customers and at the same time useful to the company as a tool for control?

In contention with Hart's guidelines,[72] which are intensively discussed in the literature, the essential requirements for an effective service guarantee will be identified – considering the quality demands that customers place on the

69 Laumann et al. (2002), p. 292.

70 See Baumeister et al. (2005), p. 487; Bockholt and Jugelt (2005), p. 3; Laumann et al. (2002), p. 292.

71 See Bockholt and Probst (2003), p. 25.

72 See Hart (1988), pp. 55 et seq.

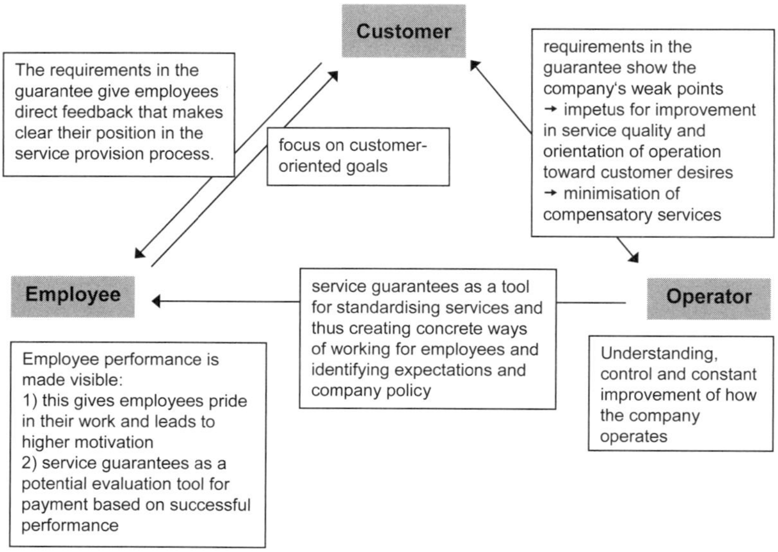

Figure 2.6 How service guarantees work

Source: author's depiction based on FGSV (2004), p. 123.

service of transport. Additionally, the previous concepts and experiences of service guarantees for public transport in Germany will be presented.

2.4.4 Formulation of Service Guarantees

2.4.4.1 Main aspects In order for a service guarantee to fulfil its function of increasing and/or restoring customer satisfaction, the guaranteed performance qualities should refer to the core aspects of the service. The standards that are defined should be important to the customer. In doing so, the guarantee should not be so mildly formulated that it will never be applied; this introduces the risk of the company sending the signal that it mistrusts some of its capabilities and thus losing its credibility.[73]

The question arises as to what aspects a service guarantee in public transport should address in order to be attractive to customers and to make sense for the company.

Customers evaluate the overall quality of their trip, which is composed of several phases which are independent of one another.[74] There is thus a failing from the customer's perspective if the performance promised in the company's travel decisions is not maintained. This includes problems in the pre-travel

73 See Hart (1988), p. 55; Laumann et al. (2002), p. 293; Wirtz (1998), p. 840.
74 See Meffert (2000), p. 7

phase such as wrong or insufficient information, deficiencies in the travel phase in the form of delays or cancellation, as well as unsatisfactory replies to complaints in the post-travel phase.[75]

Haas and Störmer[76] state that 'the consumption of services is the consumption of time'. Thus the evaluation of transport services is decisively influenced by time-based criteria. This includes travel time and the time taken to move from the departure point to the destination, as well as punctuality and arrival according to the timetable.

A journey with public transport often consists of several separate stages. Due to diversified travel routes, not all routes can be served by direct services. In order to make necessary transfers as time-efficient and comfortable for customers as possible, punctual trips must be ensured and transfers coordinated with one another. *Ensuring the network of transfers* is thus essential in public transport. *Punctuality, direct connections, schedule synchronization* and/or seamless transferring are among the basic tasks of public transport.[77]

The factors involved in punctuality thus influence the customer's perception of quality in terms of schedule synchronization. For this reason, mobility guarantees should be provided to assure these expressly stated customer expectations. Such guarantees of punctuality and connections dominate the German transport market as well (see Table 2.1). It can thus be said that transport companies' guarantee concepts refer to the main aspects of service provision.

Table 2.1 Forms of guarantee used by the transport associations/companies in Germany

Punctuality guarantee	46 (90.2%)
Connection guarantee	20 (39.2%)
Cleanliness guarantee	21 (41.2%)
Other guarantees	7 (13.7%)

Source: author's survey.

The individual mobility guarantees, however, differ in their *formulation*. For example, companies define various *minimum delays* in their punctuality guarantees which must exist in order to make a claim valid. In this, a threshold of 20 minutes is usually applied, irrespective of whether it is regional or urban

75 See Schiefelbusch (2004), p. 319.
76 Haas and Störmer (1999), p. 122.
77 See Topp (2006a), p. 25; FGSV (2004), p. 12.

transport. Nonetheless, some urban providers[78] define more narrow time-frames of 10 or 15 minutes, which shows that there is certainly leeway for the definition of low thresholds. The Northern Hesse Transport Association, Nordhessische Verkehrsverbund (NVV) sets the highest requirement with its offer of a five-minute guarantee for transport.

Furthermore, in the context of punctuality guarantees, a distinction must be made of whether the guarantee addresses *departure or arrival delays*. Companies define, on the one hand, that compensation is valid if 'the destination is reached with a delay of more than 20 minutes' (arrival delay).[79] In such formulations, the entire travel network, that is delayed arrival, includes all connections. On the other hand, a guarantee will sometimes only be valid if there are 'delays of more than 20 minutes for buses and trains at the initial station' (departure delay).[80] If the delay occurs in the course of travel, the customer has no claim for compensation. However, passengers who wait for a late transport vehicle can make a claim for compensation. From the customers' perspective, these regulations seem unfair.

Furthermore, in the formulation of a punctuality guarantee, the defined *thresholds* must be considered. In cases of cancellations, customers using the next regular departure have to wait anyway (for example, 30 minutes). If the guarantee conditions also stipulate that a minimum delay of (in this case) half an hour is required, there is already a cause for doubt and/or dispute as to the extent of the customer's claim to compensation.[81]

The attractiveness of a connection guarantee is influenced by how many *buffers* the company plans *between the arrival of one and the departure of another route.* Ensuring tight connections or connections between various means of transport (for example, train and bus), as the WVG Group offers,[82] is especially attractive to customers. Moreover, the extent of *time limitations for the connection guarantee* is decisive. Thus around two thirds of all *connection* guarantees in Germany refer to assuring connections in evening and night hours. The offer makes sense for these times because the cycle frequency

---

78 E.g. HVG (10 min.), VAG Freiburg (10 min.), BSAG (15 min.), traffiQ (15 min.)

79 E.g. punctuality guarantees for regional trains in Schleswig-Holstein or the BVG.

80 E.g. punctuality guarantees from the transport associations AVV, VRR and VRS.

81 See Jochim (2007), p. 97.

82 The bus companies in the WVG Group (North-Rhine Westphalia) provide connection guarantees between rail and bus. It is thus ensured that the bus arrives at the train station in time for the regional train, leaving a few minutes later, can be reached. The same is true in reverse: it is promised that the connection between train and bus will function as well. The buffers here are usually under 10 minutes (see e.g. timetables for the regional bus line R19. http://www.rvm-online.de/busangebot/busindex.html, accessed February 2007).

diminishes and so missing a connection means extra-long waiting times, and is thus especially frustrating. On the other hand, evening and night travel covers a limited part of travel demand.[83] The guaranteed connections therefore only apply to a comparatively small portion of the customers.

According to the Eurobarometer studies, alongside the central time aspects (punctuality and ensured connections), the following criteria influence the passenger's perception of quality:[84]

- accessibility (for example, route and station frequency, accessibility for people with limited mobility);
- availability of *information* (for example, real-time information in cases of service disturbance);
- *safety* (for example, presence of personnel, video surveillance, flawless technical condition of vehicles);
- *comfort* (for example, cleanliness, availability of seating); and
- customer service (for example, friendliness of personnel, complaint management).

Even if it is actually not possible to integrate all of the named aspects into a service guarantee, it can be said that some criteria have only been used seldomly if at all by transport companies. While an assurance of cleanliness in stations and in vehicles (comfort) in the form of cleanliness guarantees is offered by numerous transport service providers at least in Germany (see Table 2.1), it is rare to find guarantees for the exclusive usage of low-floor vehicles (accessibility), processing passengers' concerns within determined time-frames (customer service), or climate control in vehicles (comfort).[85] The fact that some companies do guarantee these aspects and offer compensation for non-compliance shows that it is possible to broaden guarantees to include additional criteria.

2.4.4.2 Significant compensation Alongside reference to the main aspects in the formulation of a service guarantee, it must also be provided that if a promise is not kept, the customer receives adequate compensation to incite them to make a claim. At the same time, a company can show the customer

83 According to the national travel survey 'Mobilität in Deutschland 2002', after 7:00 pm. only 10 per cent of daily LPT journeys are made (see BMVBW (2004), p. 86).

84 See TNS Opinion and Social; European Commission DG SANCO (2005), TNS Opinion and Social; European Commission DG MARKT (2005) European Commission (2004b), OPTEM; EU Commission, DG Health and Consumer Protection (2003) and EUSG Konsortium (2006), pp. 83 et seq.

85 These individual cases are listed in Table 2.1 under 'other guarantees'.

with significant compensatory services[86] how important their satisfaction is to them.[87]

If trains do not arrive, delays occur, or connections are missed, *immediate measures* should be taken that ensure direct continuation of travel as opposed to *retroactive compensatory services*.[88] Thus continuation of travel with a taxi is certainly the most attractive solution for the customer.

In Germany, two thirds of transport service providers offer (partial) payment of taxi costs in cases of delay. If a connection is missed, 95 per cent of transport providers pay for a customer's *taxi trip*. The respective limits of compensation vary between €5 and €30; sometimes the cost of the entire trip to the destination is covered by the transport company. The most convenient solution for the customer is the settling of the resulting costs between the taxi and transport companies because then the customer does not have to make an advance payment.

If there are limits defined for travel cost compensation, which seems to make sense to protect the operator against abuse, these limits must be chosen such that the amount can cover a reasonable distance. If one considers, for example, the taxi costs in Stuttgart,[89] €10 only covers a distance of about 5 km. According to the study 'Mobilität in Deutschland 2002',[90] however, only every third LPT journey ends after 5 km. Thus in most cases this sum can actually only be a contribution to the resulting costs. With €30, in contrast, a taxi trip of almost 20 km is possible. This would cover 75 per cent of each day's LPT trips[91] such that this sum would comprise in most cases the entire journey without financial contribution from the customer.

Instead of offering financial compensation for a taxi, around one quarter of transport providers provide retroactive compensation by *reimbursing the fare*. In local transport this often means sums of under €5. It is questionable whether this sum is impetus enough for a customer to take the trouble to make a compensation claim. In addition, most service providers that reimburse customers for poor service with tickets do not cover the cost of sending the guarantee claim. In making a claim for the cost of a simple trip, this creates costs that are indeed significant in relation to the compensation given. For example, if a customer receives a 24-hour ticket worth €4.60 from

86 In the following sections, 'significant compensation' is used to refer to compensation measures which can be considered adequate in proportion to the service malfunction and damage experienced by the customer.

87 See Hart (1988), p. 56.

88 See Schiefelbusch (2004), p. 320.

89 Stuttgart taxi fares: the base fare for a taxi is €2.50 plus 1.50 per km travelled (up to 4 km travelled), or €1.40 per travelled km (starting with 4 km travelled) http://www.taxi-stuttgart.de/taxi/taristgt.htm (accessed July 2006)

90 See BMVBW (2004), p. 104.

91 See BMVBW (2004), p. 104.

PaderSprinter as a claim on their guarantee, the postage costs for sending the form already account for 12 per cent of the compensation. At the same time, a free ticket has the disadvantage of not being useful to owners of long-term tickets. Even though long-term customers comprise the largest part of transport companies' customers,[92] a minority of transport companies that reimburse their passengers with tickets or coupons for their next trip have regulations for owners of long-term tickets such as offering discounts on season tickets.

In contrast, taxi regulations are good for all customers. Some companies are aware of the importance of their regular customers and offer owners of long-term tickets a higher contribution to taxi costs.[93]

For claims on cleanliness guarantees, all transport providers using this type of guarantee offer to cover cleaning costs completely. For the passenger, this means professional cleaning of clothing, which is often necessary to deal with tough soiling from graffiti or chewing gum on seats. At the same time, the company receives a voucher along with the receipt with which it can balance the guarantee claim.

2.4.4.3 Conditions for validity? The most debated requirement for a good service guarantee that Hart[94] identifies is the claim of providing an unconditional guarantee, that is a service guarantee that has no exceptions. Laumann, Röhrleef and Sauer[95] make a similar argument in stating that service guarantees lose more credibility with each exclusion criterion, creating a potential that the customer will become suspicious that it is not a real quality promise but merely a means of advertising. Conversely, studies have shown that customers do not necessarily rate unconditional guarantees as more attractive. The customer has to be sure that the company is actually in the position to deliver the defined standards fully. If the level of the offering seems to be too high, it makes the company look untrustworthy.[96] Stockholm local transport customers rate an unconditional guarantee as suspect as well. Public transport's bad reputation makes it untrustworthy and the lack of conditions leaves doubts. Instead, passengers prefer a guarantee with clear and precise conditions.[97]

The difficulty thus arises of keeping a guarantee's *validity conditions* as minimal as possible, on the one hand, in order to avoid making customers

92 In 2005, three quarters of all passengers within VVS were season-ticket holders. http://www.domino1.stuttgart.de/web/ksd/KSDredsystem.nsf/1622b1049bb045641256a15005a9b56/ac5588f16ec0b077c12570a500478d28?OpenDocument (zuletzt February 2007).

93 E.g. the transport associations VRR and VRS and their member companies.

94 See Hart (1988), p. 55.

95 See Laumann et al. (2002), p. 293.

96 See Wirtz (1998), p. 840.

97 See Björlin-Lidén (2004), Paper III, p. 344.

suspicious that it is only for advertisement effect and, on the other hand, providing the customers with trust in the service guarantee.

In public transport, the individual transport companies only comprise the user interface of a total system whose operation can only be partially influenced by the individual companies because of the fact that there are numerous external and internal influences on public transport. A transport company operates in a complex environment that is characterized by numerous points of contact with its surroundings. In addition, a diverse range of stakeholders are involved in the provision of transport services such as infrastructure operators. If irregularities occur in the provision of service, it is irrelevant to customers who is responsible. They want an integrated travel network. For them, the overall quality of all phases of their travel is the decisive element.[98]

Nevertheless, almost all transport providers in Germany only offer compensation in their service guarantees if they were responsible for the poor service. Reimbursement is usually not valid if the delay or cancellation was caused by illegal parking, strike, traffic, extreme weather or an accident. These exceptions defined by the transport providers are understandable because it is not possible in their legal position to hold the originators of delays accountable. For example, an illegal parker on the street tram tracks is not liable for the additional expenditure that the transport company accrues as a result of the delays.[99] However, about one third of DB AG's delays are caused by external influences.[100] On the other hand, exclusion conditions like these make it hard for the customer to judge whether the cost of a taxi will be reimbursed in any given case of delay. In addition, 'the transport company's liability [should not] depend on the transport company's fault. What is decisive for the customer is whether or not he is transported in a timely manner'.[101] If the passenger makes an advance payment and the company refuses to cover the costs due to uncontrollable conditions, the customer is then indeed unsatisfied. If the customer was thoroughly informed of the service guarantee and its conditions as well as being provided with sufficient information about the cause of disturbance before beginning their trip, then dissatisfaction can be avoided.

Nevertheless, the most attractive solution from the customer's point of view is that offered by the Nord Ostsee Bahn rail company and the Bochholt city bus, which provide compensation regardless of the cause of the delay. In addition, the example of the Dutch train Nederlandse Spoorwegen shows that

---

98 See Laumann et al. (2002), p. 293; Meffert (2000), pp. 7 et seq.; Schiefelbusch (2004), p. 319.

99 See FGSV (2004), p. 132.

100 DB AG Chairman Hartmut Mehdorn in an interview with the Stuttgarter Zeitung on 23 January 2003. http://www.fahrgast-rechte.de/Ubersicht/Politik/StuttgZtg/stuttgztg.html (accessed February 2007).

101 Pro Bahn (2003), p. 16.

service guarantees with very limited restrictions are applicable in national rail transport as well.

2.4.4.4 Precise formulation of guarantees and versatile communication So that customers know precisely what they can expect and employees know what is expected of them, the guarantee should be formulated *simply* and *precisely*. A clear promise, for example, is 'answer within three days', instead of 'answer as quickly as possible'. Also, a customer guarantee is only effective if the company receives comprehensive feedback about the service guarantee. Only then can an awareness of deficient operations be derived, and at the same time customer satisfaction be increased through compensatory services. It is thus necessary to communicate an easily understood and easily conveyed guarantee *in a versatile way*.[102] 'The guarantee has to ... run like a "red thread" through all components of the user interface.'[103]

If one looks at the media that transport companies use to publicize their service guarantees, it becomes clear that those with a broad scope such as radio, television, and print media are largely only used at the time that the customer guarantee is introduced. Permanent publication only takes place on the homepages of transport companies and transport authorities, or posters and flyers at stations. Thus, they predominantly only reach heavy users. Proactive strategies such as announcements on platforms or on-board are likewise rarely used. It is questionable whether occasional customers or new customers can be reached with such limited modes of communication.

The degree to which customers are informed about service guarantees can be seen in the level of awareness about the offerings among passengers. However, investigation of the transport companies in Germany has shown that only a small fraction of transport providers even ascertain the degree to which customers are aware of their service guarantees. The Berliner Verkehrsbetriebe (BVG) know that 40 per cent of their *passengers* are aware of their customer guarantee.[104] In comparison, 80 per cent of the *population* of Stockholm are aware of the guarantee regulations of their local transport providers.[105] These figures substantiate the fact that there is significant room for improvement in the level awareness about guarantee offerings and that the current communication of guarantee services is apparently not exhaustive.

The number of incoming claims on the guarantee can also provide information on how aware passengers are about guarantee services. A survey of transport providers has shown that in 2005, around three quarters of providers received fewer than 100 claims on their guarantees. If one considers

102 See Hart (1988), p. 56; Wirtz (1998), p. 840.

103 Bockholt and Jugelt (2005), p. 2.

104 See Ekoloff and Fulton (2005).

105 This figure was determined directly after the introduction of the guarantee and the accompanying marketing publicity, see Björlin-Lidén (2005).

that in this time period transport companies transported many millions of passengers, the feedback that companies received through their guarantees can be classified as extremely marginal.

Over half of the companies questioned say that the costs of guarantees are below their expectations, at least. Companies explain this as resulting from the low level of awareness. One quarter of the companies cite the fact that the quality standards for punctuality and guaranteed connections were met more successfully than initially expected. This raises the question of whether the guarantees are too 'softly' formulated, that is leaving definite room for a tightening of regulations – for instance, that the guarantee becomes effective starting with 10 minutes of delay (see Section 2.4.4.1). They also say that although customers may have the opportunity, they do not use the guarantees. Based on this, one could draw the conclusion that the compensatory services are not attractive enough. It is thus possible that numerous customers refrain from spending the time and money necessary for making a claim when the return service is a ticket worth less than €5 or when there is a danger that their taxi costs will not be reimbursed because the company claims that the cause was out of their control (see Sections 2.4.4.3 and 2.4.4.4).

Another result of the low level of feedback is that there is only a weak pressure to optimize the services.

2.4.4.5 Easy claim processes for compensatory services A customer who is already dissatisfied because of poor service quality should not have to overcome numerous obstacles in order to make a claim on the guarantee. Instead, making a claim for compensation should be made as easy as possible. This includes providing a wide variety of contact options such as mail, e-mail and telephone. Also, the burden of proof for deficient services should not be placed on the customer.[106] As Wirtz states: 'service recovery by means of a guarantee cannot be achieved through bureaucratic documentation'.[107]

In the customer's view, the *compensatory procedure* should therefore have *as uncomplicated a design as possible*. For example, payment should be made according to the customer's choice (bank transfer, coupon, cash, and so on) and soon after the incident.[108]

In cases of poor service from a company, it is necessary to have information about the incident such as the time, place or mode of transport used in order to document the event. Ready-made forms make it easier for passengers to document incidents. The majority of transport providers who were asked provide forms as downloads on their company homepage or place them in customer centres. Many companies also offer to send their forms upon request, or accept submissions without a form. The availability of guarantee forms is

106 See Hart (1998), p. 56; Wirtz (1998), p. 840.
107 See Wirtz (1998), p. 840.
108 See Wirtz (1998); p. 840.

thus certainly adequate, even if only a few providers use the full spectrum of options such as additional printing of forms in customer newsletters and disbursement in vehicles and stations.

In order for compensatory services to be provided, transport companies demand that certification of resulting costs and used tickets be sent to them, which makes sense for providing protection against misuse of the regulations. There should be no additional expenditure expected of customers for submitting forms and other documents, such as postal charges for small reimbursement sums (see Section 2.4.4.2).

The attractiveness of a service guarantee is influenced by the *deadlines* for sending claims as well as by the processing times. Among the companies that offer service guarantees in Germany, the average time span that customers have to observe for claims to be valid is 12.6 days, which can be seen as adequate. Companies promise customers an answer within an average of 8.2 days. Thus companies for the most part impose tighter deadlines on themselves for the processing of claims than they do for customers. On the other hand, it makes regulations questionable if they demand that customers submit their claims within a certain period but make no such demands on internal processing times, as do 20 per cent of companies asked.

In summary, it can be said that guarantee claims procedures can for the most part be considered uncomplicated. To date, only three providers have explored how satisfied customers actually are with offered service guarantees. Customers tend to regard the complexity of submitting a claim as reasonable.

2.4.5 Deficiencies in the Implementation of Service Guarantees

Service marketing literature and the writing in this chapter concentrate principally on the advantages of service guarantees. In these times of increasing competition in the transport market, these voluntary commitments are thus seen as potential tools for strengthening competitiveness, optimizing operations, and increasing customer satisfaction. Potential undesirable results of a guarantee can include:[109]

1. negative impact on the employees, who might become unmotivated through continual negative feedback coming from guarantee claims;
2. the danger that customers see the offering of a guarantee as a confession that there are potential problems in using the service, thus raising consumers' perceived risk;
3. the danger of misuse of the regulations; or
4. significant financial burdens for the company, which in public transport necessitate increased government subsidies.

109 See Laumann et al. (2002), p. 293; Wirtz (1998), pp. 842 et seq.

These factors should indeed be considered in the decision-making process, but would however not negate the potential usefulness of this tool. In addition, research concerning the last two points has shown that expenditures for guarantees has remained significantly below expectations. The Dutch Railways NS have made the same report – despite their extensive guarantee – with only around five complaints per delay.[110] Furthermore, the companies asked view misuse of guarantees to be extremely low.

Many companies have set for themselves the challenges of increasing customer-orientation and have thus recognized the potential of customer guarantees. In 2006 alone, two transport associations and five transport companies introduced new guarantee services. In this way, transport service providers have recognized the capabilities of service guarantees to reference the *main aspects of their service* and, in guarantee claims, to *facilitate the receipt of compensatory services*. However, in order to achieve the intended effect, that is to utilize the potential of service guarantees, from the current perspective there is still a significant *need for optimization* in terms of:

- *design* (for example, the definition of specific parameters such as guaranteeing connections at all times of day, separate regulations for people who have long-term tickets, or relaxing exclusion criteria);
- *implementation* (for example, intensification of communication and implementation of studies about the acceptance of and satisfaction with service guarantees); and
- enforcement (the integration of service guarantees in transport contracts, codification of guarantees in companies' terms of carriage).

The deficiencies in the usage of guarantees described here can be demonstrated in the majority of transport providers in Europe in a Europe-wide comparison.[111]

Furthermore, in order to classify the results, it must be said that the introduction of services guarantees onto the transport market addresses a development that is still new. Excepting the introduction of the first guarantees at the end of the 1990s, the process of establishing guarantees saw its high point at the turn of the century. The experiences that companies have had to date should in no way be seen as exhaustive; development in this area remains largely open.

110 See Verbraucherzentrale NRW (2004).

111 See EUSG-Consortium (2006), pp. 130 et seq.

2.5 Neutral Alternative Dispute Settlement – Concepts for Ombudspersons and Arbitration Boards

Martin Schiefelbusch[112]

The following section presents approaches to *alternative dispute resolution* (ADR) in public transport (PT), which falls between company complaint management systems and passengers' rights. These approaches do indeed involve references to the nature of the legal relationship between passenger and company (see Section 2.6). However, such an alternative conflict resolution service is possible and sensible in its own right, independent of developing the passengers' rights in a strict sense. It should not be understood as merely a way of dealing with claims against the company, but as an opportunity to improve the relationship between customer and provider in general.

This section thus gives an overview of the fundamental issues in the formulation of ADR arrangements, and finally presents some of the widely ranging concepts from around Europe.[113]

2.5.1 Usage and Features of Alternative Dispute Resolution

Frustration with others and conflict have been part of the human experience since time immemorial. In order to deal with this, civilized societies have created three fundamental strategies: consensus building through negotiations among the conflicting parties themselves, precipitating a solution by including a third party and cutting off further contact with one another (as much as the situation allows). If a third party is brought in to resolve the dispute, there are three fundamental concepts to be distinguished:[114]

1. In the *legal process* both parties involved in the conflict submit to the authority of the court, which judges according to the parameters provided by the legal terms established. It is central to its legitimation that it can draw upon the norms decided upon by the legislature and thus 'in the name of the people'.
2. The *decree* as well is based on the legal parameters. An authority requires the company to refrain from an activity or compensate for illegitimate actions.
3. In *arbitration* a third party is engaged to develop a proposal for resolution. This proposal is generally not binding for both parties as is

112 For their commentaries and suggestions, I would like to thank Mr Christian Moritz and Mr Stefan Böhm, Europäisches Verbraucherzentrum Deutschland, Kehl as well as the employees in the various ADR bodies who were interviewed.

113 Due to spatial constraints, not all aspects can be addressed here. See also Schiefelbusch (2007a).

114 See Heger (2003); Maiwald (2004).

a verdict, but can only become effective through amicable acceptance by both parties. With regard to bindingness, different variants are to be found in terms of the degree of involvement for both parties, and the courses of action in the case of failure.
4. In *mediation* the formative role of the third party recedes; their role is essentially to support the conflicting parties in the creation of a common resolution that they both accept. Both forms of extra-judicial dispute resolution (3. and 4.), because of the lack of formal legitimation, thus necessitate interest in a solution from both sides and (remaining) willingness to cooperate.

In the following text, term 'arbitration' will be used comprehensively – according to English linguistic usage – even though arbitration is only one of the many forms of extra-judicial dispute resolution.

Extra-judicial practices (3. and 4.) are usually less a matter of public focus than of legal practice, but they can provide an important contribution to dispute resolution. They offer several practical as well as conceptual advantages:[115]

- increased cost-effectiveness and efficiency;
- disburdening of justice;
- an agreement that sets no precedent (it is only effective for the individual case);
- more freedom in forming the content of the agreement; and
- the opportunity to find an amicable resolution that allows for the continuation of the (business) relationship.

Because of its product characteristics and market structure (see Chapter 1), public transport has a special need for such a service. In view of the passenger's legal position, there is a lack of legal foundations and/or empowerment to effectuate a legal review of disputed issues. The often low financial value of disputes (which can nevertheless lead to significant frustration) make this additionally unattractive.

2.5.2 Basic Possibilities for the Formation of Arbitration Organizations

In making comparisons, it becomes apparent that no two arbitration boards are the same in all aspects. Thus, before examining the examples, it makes sense to explore a few conceptual issues that are decisive in the development of such a facility:

115 See, e.g., Europäisches Verbraucherzentrum Deutschland (European Consumer Centre Germany) (n.d.), p. 6; Seneviratne (2000); Tonner (2007).

- *Field of responsibility*: Should the board be responsible for the transport field alone, or (comparable to a court) should transport issues be dealt with in cooperation with a general organization?
- *Field of activity:* What kind of conflicts should the board handle?
- *Support*: Is a legal foundation necessary, or is the companies' support enough?
- *Accountability*: How binding should the board's decisions be for customers and companies?
- *Arbitration practices*: Who is concretely responsible for intermediation and how does this work?
- *Positioning*: Does the board see itself as neutral or does it represent the passengers (comparable with an advocate, see Figure 2.7) against the company?

Further practical issues concern financing, personnel and qualifications, as well as documentation and public relations.

The various design possibilities are also apparent in the names that are used. Common terms are 'arbitration board', 'adjudication board' and 'ombudsperson'. The Swedish term 'ombuds(person or -board)' has found widespread international use. According to Hopp,[116] this position is connected to the processing of individual cases. He emphasizes that the neutrality and autonomy of this position are stronger than in the German understanding of the term *Schlichtung* (arbitration).[117] An ombudsperson is seen rather as an attendant who focuses on the concerns of specific issues or groups that are not adequately represented in the general political system.

The difficulty of framing the various models is also increased by the fact that these terms are not necessarily consistently used. It is, however, possible to identify two extreme cases which, according to their legal processes, can be termed the advocate model and the judge model. Figure 2.7 contrasts these in terms of their most important criteria. This addresses ideal models that in reality – except when legal processes intervene – are not to be found. The similarity of both models remains, however, a desire to spare 'hard' legal proceedings and solutions that might lead to an unnameable solution that might harm the future relationship of those involved. They also both strive to provide a process that is as low threshold and transparent as possible (see following section).

2.5.3 Quality Criteria for Arbitration

In view of the diversity of possibilities for forming an arbitration board, it is especially important to determine the (quality) criteria that every organization

116 Hopp (1992).

117 See Seneviratne (2000).

ADVOCATE		JUDGE
Represents the client (is partial)		Weighs arguments from both sides (offers neutrality)
Negotiating basis is one's own position without commitment to the opposition		Makes a binding/final decisions 'for both parties'
Consensus building through dialogue with the company ('external')		Consensus building through in-house consideration, possibly through boards with equal representation
Combination with general and/or political goals (which extend beyond the individual case) is possible	◀ ▶	Appraisal of each case is individual, other decisions used only for orientation
The public can be involved as an 'ally'		Dealings are initially confidential, but reporting is part of the transparency principle
Example: Passenger Focus (Great Britain)		Example: SGC (The Netherlands)

Figure 2.7 Comparison of 'advocate' and 'judge' models for ADR

Source: author's depiction.

should fulfil. The EU Commission adopted a recommendation in 1998 (98/257/CE) which posits seven fundamental principles:

1. *Autonomy*: a fundamental principle is that the board act impartially. This is ensured by the *expertise* of the appointed person(s), their *appointment for an adequate amount of time*, and significant *barriers to their dismissal.* Furthermore, they *cannot* have been previously *employed by the company* whose activities are addressed by the board. In collegial arbitration boards (consisting of several people), providers and consumers should be *equally represented*, if necessary by a leader who is neutral in the aforementioned sense.
2. *Transparency*: the board communicates to the consumers its field of activity, its mode of operation, and the possible results of their practices in a suitable form and informs the public about its work in an annual report.
3. *Procedural dialogue methods*: the parties must have the possibility to represent their position and answer to the comments of the opposing side.
4. *Efficiency*: the board should provide *efficient processing* with clear deadlines. It must be able to procure the necessary information from both sides. Secondly, low-threshold access from the consumer's perspective

must be ensured as a central aspect of arbitration. This includes *inexpensive or free processing* without requiring the intervention of a legal representative.

5. *Legality*: the board may not ignore the mandatory consumer protection laws of its respective country. Its decision must be justified and communicated in writing.
6. *Freedom of action*: the decision can only be binding if it was previously announced as such, and if both parties have agreed.
7. *Representation*: the consumer must at all times be able to have him- or herself represented by a third party, or to call upon one.

Beyond the criteria named in the Commission's recommendation, those involved in the conflict – especially the company – must be prepared to cooperate with the arbitration board. Cases should be individually examined and the position should be comprehensibly justified. Contact persons are to be named and stipulated deadlines maintained.[118] In order for the work to be afforded acceptance, it can make sense to have an advisory board composed of consumer and provider representatives or a commission with equal representation.

These basic principles are reflected in the arbitration board's work and/or statutes. However, not all boards work according to the rules defined in the EU recommendation. In Germany, for instance, this is currently often (still) not the case because the stakeholders – especially companies and their associations and organizations (for example, Chambers of Commerce) – have been accorded more leeway in the design of arbitration schemes. German boards thus rely on voluntary agreements and commitments without a legal context. Thus, the provider's concepts can have a larger influence on the design of an arbitration board, even if they see themselves as neutral in practice and draw their legitimation from a mutually recognized neutrality.

2.5.4 Arbitration Boards in Public Transport

Belgium provides the most comprehensive ADR offerings in Europe for public transport. A total of four ombuds-services deal with PT. Their work is based on national or regional legal foundations, which determine their practices and fields of activity, and guarantee their rights in relation to service providers (for example access to records and examination of witnesses).

Every Belgian ombudsboard publishes annual reports, which are particularly oriented toward people involved in politics. The attention garnered in this manner is seen as important for their work. The reports sometimes include general recommendations for solutions to problem areas. Furthermore, in

118 See Lobmeyer and Zandke-Schaffhäuser (2007).

order to apply political pressure, regional boards can present topics to the respective ministers responsible and put forth legislative proposals.

In *France*, to date there has only been a so-called *médiateur* appointed at the two largest transport companies, the national railway SNCF and the Paris RATP. Until 2006, requests for arbitration could only be submitted through one of the 17 consumer organizations that existed at the time; since then requests can be submitted directly. The organizations, including the passengers' organization FNAUT, conduct a preliminary examination of the case and then represent the customer against the company, comparable to a legal advocate. The *médiateurs* are employees of the company, but work independently. The consumers' organizations have a right to have a say in the choice of who fills these positions.

For a long time, when passengers in *Germany* had conflicts with transport companies they were limited to their own negotiating skills and the political means described in Section 4.2 – and in many cases they still are. Since the beginning of the twenty-first century, however, various ADR bodies for certain regional and modal areas have been created (see Table 2.2).

The first arbitration facility was created in 2001 in North Rhine-Westphalia, the Schlichtungsstelle Nahverkehr (Local Transport Arbitration Board), in the context of the country[119] government's 'Masterplan Qualität' ('Quality Masterplan') for public transport.[120] It functions as an 'independent body for disputes between customers and transport companies'.[121] It also does comprehensive public relations that significantly exceeds annual reports and casework, making it eager to give voice to fundamental problems as well. Examples include the publication in recent years of an arbitration index meant to offer indications for reasonable compensation, and the collection of delay data in LPRT (*Pünktlichkeitsprotokoll*, or punctuality protocol).

The *Leipziger* Verkehrsbetriebe Leipzig Transport Services (LVB) was pursuing another concept with the institution of an *ombudsperson*. This was meant to give passengers an additional possibility to express criticism and concerns about the company. It was not important in this case whether these were dealing with first-time complaints or dissatisfied customers who had already been through customer service. The LVB tried to make the ombudsperson accessible through notices in vehicles and regular office hours, among other things. In hindsight, it becomes clear that a further advantage was that the ombudsperson could pursue even complicated problems for a long period of time. Her position was only filled for two years to ensure

119 In this book, the term 'country' is used to refer to the sub-national territorial units (*Bundesländer*) in Germany and Austria.

120 Ministry of Enterprise and Trade, Energy and Transport NRW (Ministerium für Wirtschaft und Mittelstand, Energie und Verkehr NRW) (2001).

121 http://www.schlichtungsstelle-nahverkehr.de/UNIQ117227404522277/doc12270A.html (accessed February 2007).

an ombudperson's perception of problems uninfluenced by 'routine tunnel vision'.

At the end of 2004, the *Schlichtungsstelle Mobilität* (Mobility Arbitration Board) was the third facility to be created by the Ministry of Consumer Affairs. The impetus was provided by the discussion begun in 2002 about passengers' rights that finally lead to the passing of a customer charter through DB AG. In this, DB required itself to support a 'neutral Mobility Arbitration Board initiated by the federal government that comprises several transport providers'.[122] The Board's spatial and modal field of activity was thus determined politically; however shortly thereafter the Federal Air Transport Authority appointed a further board for processing air transport disputes in terms of EC Ordinance 261/04. This fact, as well as the somewhat lacking cooperativeness of air transport companies have so far made for difficulties in practice.[123]

At about the same time, the *Bavarian Local Transport Ombudsboard* was created as a pilot project by the regional service provider associations. The managing directors of these associations act as ombudspersons (arbiters), and sometimes also work for transport companies. Their arbitration strategy is based on the close contact to companies that arises as a result of their activity, and draws essentially on informal dialogue. Public relations takes place only on a modest level; furthermore the board reserves the right not to deal with complaints oriented toward transport politics. Since the end of 2005, a very similarly structured *Local Transport Ombudsboard* has been formed in *Baden-Württemberg*.

In *Great Britain* and *Northern Ireland* there is a comprehensive system of institutions that are focused on representing consumers' interests in general, as well as in individual cases (see Sections 3.5, 4.6.5, 5.7). Accordingly, most of the organizations see from the consumers' perspective and work as watchdogs. In transport, these include:

- Passenger Focus, responsible for rail transport;
- London Travelwatch (see Section 4.6.5); and
- The General Consumers' Council of Northern Ireland.

In arbitration, they take on the role of advocate (see Figure 2.7). The process is more similar to negotiations of a contract than they are to a court case: the representative takes the case on in the customer's name and uses special contacts within the transport companies to try to find a better solution for the customer.

In rail transport, alongside the watchdogs there is the Independent Penalty Fares Appeals Service (IPFAS) which is responsible for neutrally judging

122 Press release BMVEL/Deutsche Bahn from 3 February2004.

123 See http://www.schlichtungsstelle-mobilitaet.org/52.html (accessed May 2006).

Table 2.2 Features of the arbitration boards in Germany (as of mid-2006)

Board's name	Arbitration Board for Local Transport NRW	Arbitration Board for Mobility	LVB Leipzig Ombudsperson[a]	Bavarian local transport Ombudsboard	Baden-Württemberg local transport Ombudsboard
Founded	2001	2004	2002–2004	2004	2005
Supporting organization	Verbraucherzentrale NRW	Verkehrsclub Deutschland	Leipziger Verkehrsbetriebe	VDV and BDO state associations	
Funding	NRW (until 2007) now consumer association/association of PT operators	Ministry of Consumers' Affairs (Federal)[b]			
Budget p.a. (€m)	0.2	c. 0.5	n.a.	Limited (materials costs)	
Positions[c]	3	6	1	No positions of their own	
Stance	Independent mediator representing consumers	Neutral	Independent position within the company	Affiliated with transport associations	
Formal foundations for the work	Political initiatives		Decision of the company's management	Decision of the association	
Field of activity	LPT and LPRT in North Rhine-Westphalia	Long-distance rail, bus, air and water transport in Germany	Regular transport services of LVB	LPT and LPRT in Baden-Württemberg and/ or Bavaria except for issues concerning service patterns, fares and transport politics	

Table 2.2 cont'd

Board's name	Arbitration Board for Local Transport NRW	Arbitration Board for Mobility	LVB Leipzig Ombudsperson[a]	Bavarian local transport Ombudsboard	Baden-Württemberg local transport Ombudsboard
Settlement procedure	Proposal made by the arbitration board; if applicable through negotiation with the TC, direct discussions in exceptional cases	Proposal made by the arbitration board; if applicable through negotiation with the TC	Proposal made in agreement with the affected branch	Discussion of problems with the company over the telephone is preferred	Discussion of problems with the company over the telephone are preferred, in exceptional cases direct discussions
Advisory board	No	Yes	No	Yes	No
Nature of the conclusion	Recommendation	Recommendation	Recommendation	Recommendation	Recommendation, acceptance is a voluntary commitment on the TC's part
Board's options in cases of refusal	Referral to superior bodies (publicity[d])	None	Submission to the management	None	None
Public relations	Homepage, customer magazine, press reports, displays in the Consumers' Central's information centres	Homepage, media work, announcements	Internet, notices in vehicles, press	Internet, some notices in the TC's media	To date only press releases at the beginning
Activity reports	Public	Internal, but documented online and in trade publications	Internal	None	Internal
Cases completed[e]	2,500 (2005)	c. 2,060[f]	596 (2003)[g]	c. 40 (12 Months)	Not yet clear

Table 2.2 cont'd

Notes

a Board since discontinued.
b Funding in set-up phase.
c Or equivalent hours in terms of full-time positions.
d Publicity not for individual cases.
e Including advising, not including informational inquiries, erroneous complaints, etc.
f 12-month average.
g All requests.
n.s. not specified.
TC transport company.

Source: author's depiction based on information provided by the ADR bodies.

penalty fares. The main features of their case processing are legally regulated. In contrast to the other boards represented here, IPFAS serves as a first instance for appraisal; the transport companies only issue Penalty Fare Notices.

The watchdogs cover that part of the transport market which is dominated by the principle of competition for the market. For the remaining areas, ADR offerings have been introduced in recent years:

- the Bus Appeals Body (BAB) is supported by both the CPT (Confederation of Passenger Transport) and the (bus) passengers' organization Bus Users UK and processes bus customers' disputes in cases where no other organization is responsible (that is, in Wales and in England outside of the London metropolitan area).
- In Scotland the Bus User's Complaints Tribunal provides a similar board. The essential difference to BAB is that the Tribunal has a legal foundation and is required to report to the Scottish Parliament.

Both organizations see themselves as facilitators in *individual* disputes concerning things such as delays, unfriendly personnel and other service deficiencies. Basic issues concerning planning, fares, etc. are not admitted.

In *Italy*, extra-judicial dispute settlement is offered by several institutions (see Section 5.6.6). Most important in this area are company and consumer organizations, chambers of commerce, public advocates in local authorities, as well as the companies themselves. Consumers' organizations as well offer their members support in complaints, partly also in the form of arbitration. To date, however, it has only been possible to reach a small number of agreements between companies and organizations in the transport field. Probably the only real offering for arbitration in LPT is in the Ligurian city La Spezia.

In *The Netherlands*, the Foundation for Arbitration in Consumer Issues (Stichting Geschillencomissies voor Consumentenzaken, SGC) provides a facility that comes close to the judge model. The SGC is divided into 37 (as of September 2005) departmentally divided arbitration commissions, which were instituted according to a contract between the SGC and Dutch trade associations. In 1992, the first commission in PT began activity in rail transport; meanwhile other commissions have emerged in other forms of PT and in taxi transport.

The main arbitration board in *Sweden* is the State General Claims Department (Allmänna Reklamationsnämd, ARN). The Department addresses consumer disputes in various sectors, however, only when the minimal dispute amount is reached. In transport and travel, this amount is SEK1,000 (c. €110). It is thus not possible to address many LPT issues. The process is carried out exclusively internally and is based on the mode of operation in a court.[124] A consumer organization publishes an annual blacklist – based on ARN data – of the names of companies that have not complied with the Department's arbitration proposals.

In 2001 in *Switzerland*, the Public Transport Ombudsboard was founded based on an initiative by the trade association VöV. Three attorneys work out of their own offices as ombudspersons on a part time basis. Cases can be submitted to the ombudspersons and after an initial assessment are sent to the appropriate transport company with a request for response. In more difficult cases, the ombudsperson also leads direct discussions between the parties. As a last resort, the ombudsboard can express a recommendation of how the case should be resolved in their view. They have, however, no decisional authority over the company.

The structure of the board is interesting for a number of reasons: the VöV's sponsorship is important for its acceptance with TCs, the implementation of the actual arbitration work through external attorneys is, in contrast, well-suited to expressing the neutrality of the process. Furthermore, it is thus possible to react flexibly to fluctuations in the volume and complexity of cases.

2.5.5 Volume of Cases and Results of Arbitration

Due to the varying basic conditions, the volume of cases can only be meaningfully interpreted as an indicator to a limited degree. Despite this fact, the following is meant to provide a short overview of the extent of arbitration work. Table 2.3 contains the arbitration cases processed in recent years by various boards – among organizations that comprise several fields, only those addressing PT are counted. In order to be useful as an indicator, this value should be considered according to the size of the respective transport market. Due to the highly various geographic and modal layout of the respective areas

124 For more info about ARN see Johansson (2007).

Table 2.3 Number of arbitration cases in public transport

Country	Board	Year	Cases (completed arbitration)	Cases per million trips
D	Schlichtungsstelle Nahverkehr	2004	852	0.36
D	LVB Ombudsperson	2003	596[a]	n.m.
D	Schlichtungsstelle Mobilität[b]	2005	c. 510[j]	2.01
D	Bavarian Ombudsboard	2006	40	n.s.
D	Baden-Württemberg Ombudsboard		No data yet available	
B	SNCB Ombuds-service	2004	c. 1,200[c]	6.73
B	Vlaamse Ombuds-service	2004	c. 100	0.24
B	Médiateur Région Wallonne	2004	14	0.07
B	Service de Médiation STIB Brüssel	2004	72	0.30
F	Médiation SNCF	2004	422	0.48
F	Médiation RATP	2004	62	0.02
GB	Passenger Focus	2004/5	4,356[d]	4.22
GB	LondonTravelWatch	2004/5	587[e]	0.20
GB	General Consumers' Council Northern Ireland	2005/6	63[f]	0.84
GB	Bus Appeals Body	2004/5	140	0.06
GB	BUCT Scotland	2004	126	0.27
GB	IPFAS	2006	c. 30,000[g]	29.13
NL	SGC Openbaar Vervoer	2005	38	0.05
S	ARN	2005	c. 310[h]	n.m.
CH	Ombudsboard for PT	2004	58[i]	0,04

Notes

n.s. Not specified.
n.m. Not measured, i.e. not suitable as a reference value.
a All customer statements acc. the activity report.
b Estimate for 12 months.
c Total number of entries 2,378, of which c. 50 per cent were first-time complaints

d Including complaints about the railways in the London metropolitan area, which are processed by LondonTravelWatch.

e Not including railways (see Passenger Focus).

f Values for bus and train.

g All complaints, no distinction possible between completed processing and other settlements. Cases refer to the total volume of train transport, because some business use another arbitration board, this leads to a somewhat lower value per million trips.

h According to disclosures from ARN c. one third of the 969 cases in the 'travel' field are air/train/bus transport not including package tours.

i Not including cases in the canton of Zurich where there is a separate ombudsboard for public administration

j 12 month average.

Sources: case numbers: annual reports and disclosures from the boards – volume of transport: established by national transport statistics and disclosures from companies, in part by estimates).

of operation, there are major differences that do not necessarily correlate with the determinable scope of the absolute number of cases.

The British IPFAS comprises by far the largest case volume despite its narrow field of activity because this organization is contacted for the most part directly – without previous attempts for review by the transport companies.

In international comparison, Belgian and British train transport accumulate an unusually high case volume. The German Schlichtungsstelle Mobilität holds a middle position in the overall list, the rest of the organizations are, however, all significantly under one case per million trips.

In explaining these differences, a number of factors come into question; alongside the previously named influences these include:

- a company's customer service;
- cultural differences in dealing with complaints;
- requirements of the arbitration process;
- the disputed amount in comparison to the costs.

The number of arbitration cases actually processed completely can, however, only be taken as one of many indicators. The boards' reports in fact show many *additional effects*: one thing to be observed is the fact that arbitration cases are often settled or solved through additional parallel contact between customer and company before arbitration decisions are made. To some extent, the statutes explicitly provide for this. In these cases it can be supposed that the existence of an arbitration board has at least supported the settlement. Also, the boards are often contacted with *requests* that are outside of their field of activity – mainly because the case does not fall inside of their factual or spatial field of responsibility or because there has not yet been a first-time complaint against the service provider. In such cases, the boards can at least provide advice or refer the concern to someone else. These services are not contained in the above values, but could certainly comprise a multiple of the actual arbitration cases.

Such questions show the difficulty of communicating the challenges and (often indeed historically rooted) limits of arbitration work. The boards are thus called upon to further develop their *public relations* and to cooperate with companies and other boards in referring 'incorrect' requests.[125]

Even beyond the contacts and complaints recorded by the boards themselves, it seems probable that the existence of such organizations has an influence on the company and its complaint management. In this sense, one of the effects of having the possibility of such an appeals authority consists of preventing complaints from going to arbitration boards through better customer service.[126] This effect is definitively positive from the passenger's perspective, but it is also difficult to demonstrate and is not shown in any of the board's own statistics.

2.5.6 *Resources and Employment of Labour*

In accordance with the differences in case volumes and areas of activity, the arbitration boards are set up in very different ways. The following types can be observed:

- large ADR organizations such as the ombuds-services of SNCB, the Belgian and Italian regions, or ARN employ 10 to 20 people, not all of whom, however, necessarily deal with transport;
- the British watchdogs have double-digit employee counts, but only a small number are active in processing individual cases;
- mid-sized ADR organizations with two to six positions (Mobilität Boards, Schlichtungsstelle Nahverkehr, médiateur SNCF, IPFAS)
- 'combined' organizations such as the Bus Appeals Body, BUCT, SGC, consisting of a (part-time) secretariat and a regularly convening multi-person arbitration board;
- other small organizations in which individuals, sometimes supported by a secretariat, carry out arbitration full- or part-time (ombudsboards in Germany, Brussels, SNCF).

This spectrum shows that arbitration boards accommodate very different needs. Part-time regulations for arbiters and/or secretariats play an important role in this. This rebuts the argument that such organizations are not worth it

125 See Osmotherly (2003).

126 See Seneviratne (2000).

because of the low number of cases they expect. However, in the structuring of boards on a part-time basis it is important to make sure that the independence of the arbitration work is not brought into question by the arbiter's other activities.

A board's annual budget can range from a few thousand euro (the Bus Appeals Body) to the overall budget of the Swedish ARN of €2.6 million. A board's *funding* is in many cases provided from the public purse, sometimes by the trade associations, and very rarely by individual transport companies. Only in the case of the Bus Appeals Body is a passengers' organization included through the assumption of secretariat activities. In the ombudsservice Brussels, there is common sponsorship through public and company assets. (Partial) public funding can be justified by the relief that ADR provides judiciary processes.[127] Funding through service providers depends partially on legal obligation (Brussels, SNCB, IPFAS).

2.5.7 External Visibility

Alongside the previously described criteria, the degree to which the board is known to travellers has a large influence on the case volume. The board's public relations depends on the organization's legal and material possibilities, but also on how the organization sees itself. Organizations that understand themselves as customer advocates – also due to their other functions – thus take an offensive stance in public relations in that they prepare press releases about their annual reports or topical issues and take part in the transport politics discussion. In other boards as well, however, there are examples of comprehensive public relations work. Table 2.4 summarizes the instruments applied – roughly ordered according to importance.

Annual reports can be distinguished from other media in that they are oriented toward a *specialist and political readership*; it is certainly there that they are read most thoroughly.

The other instruments are intended to inform *potential users* about the offering. On-site notices seem particularly obvious, that is in vehicles and transport company correspondences. The experiences of those questioned showed, however, that if the service provider is not required to cooperate, this is often only successful to a limited degree.

In contrast, press releases require news value in order to actually appear in the newspapers. However, it has been shown that press releases about concrete cases are not helpful while arbitration processes are still in process. Generally, arbitration boards' press work is thus limited to general issues; presentations of individual cases are however – in anonymous form – often used in annual reports.

127 Tonner (2007).

Table 2.4 Arbitration boards' public relations

Medium	Prevalence
Annual report	Almost all boards examined (see Table 2.2), at times only available upon request
Internet sites	Almost all boards examined, various scope of content, in some cases are only to be found as sub-sites of the TC
Press releases	Mostly to publicize the annual report, for boards following the advocate model also address other topics
Notices in stations/vehicles	TC required to do so in GB, F, B, otherwise rare
Notices in TC's writings	TC required to do so in GB, otherwise rare
Displays, etc.	Rare (the Schlichtungsstelle Mobilität, SGC, ombuds-service SNCB)
Office hours, etc.	Rare (médiateur Wallonie, the Schlichtungsstelle Nahverkehr)

Source: author's depiction based on information provided by the ADR bodies.

2.5.8 Issues in PT Arbitration Cases

In principle, arbitration cases can address all aspects of what is known as public transport. Overall, however, the area that dominates is *customer service* in the broadest sense:

- fares, functioning and availability of ticket machines and validators, and so on;
- ticket checks and increased fares;
- passenger information;
- customer service and conduct of personnel;
- service disturbances, especially delays, cancellation, missed connections, and their consequences for the passengers.

Arbitration must thus address the concrete situation in each case, but must also not lose sight of *fundamental problems*. This may include:

- fare terms and conditions are not understandable or lead to hardship;
- badly designed ticket machines and information systems;
- poor planning of such things as capacity measurement, leading to regular over-crowding.

Another category, also an abstract one, is the actual *service planning* (timetables and routes) as well as issues of transport politics. In such cases, arbitration boards are not supported as the sole instrument, particularly because they only address a fraction of customer concerns – somewhat like the tip of the iceberg. However, along with companies' complaint management, supervisory bodies, and consumer organizations, they can contribute to revealing weak points.

Some boards, however, explicitly exclude – to various extents – issues of planning and general principle (see Table 2.2 and above descriptions). According to their nature, organizations following the advocate model are more interested in such issues.

There are also exclusion criteria drawn from the role of extra-judicial dispute resolution: often complaints are not allowed if preliminary court proceedings are underway or are possible, as well as complaints that involve personal damages. These limitations vary. Overall, these reasons for excluding cases refer to relatively rare issues.

2.5.9 Comparisons of Arbitration Concepts

Of the quality criteria contained in the EU Commission recommendation, the criterion *independence* is particularly important. It is, however, only relevant for boards that are conceived as neutral facilitators, while boards following the advocate model must, by definition, take a stance.

The EU Commission's recommendation names several qualities that are meant to guarantee independence (see above). Most of the boards examined here largely follow these specifications. However, this is questionable where arbitration organizations are situated at companies or trade organizations – such as SNCF, RATP, and the German VDV Ombudsboards. While the French médiateurs and the LVB ombudsperson are recruited from outside the company, this is not so in the other cases.

The requirement for *transparency* demands information about a board's practices and activities. Most boards fulfil these demands (see above), even though their requirements for active public relations vary greatly. The Schlichtungsstelle Nahverkehr and the Schlichtungsstelle Mobilität carry out particularly extensive public relations work. The former LVB ombudsperson as well made her service very public, but did not publish any reports.

The documentation examined did not show any reference to regulations that fundamentally prohibit making a statement during the process (*dialogue procedures*). Numerous arbitration boards explicitly arrange for inquiries with the parties. This can even be included in common proceedings. In practice, this can however take on very different forms. In the simplest case, a dialogue of this sort can take place in the context of direct arbitration talks – which, for practical reasons, cannot be offered in every case and for a long while has not been a universal offering of all boards.

The *efficiency* criterion contains several aspects: most boards work for customers free of charge, in other cases fees seem inexpensive in comparison to court proceedings, but in terms of the disputed amount they are comparatively high. It is not common to name deadlines; due to the highly differing cases this certainly seems to make sense. British offices, however, require that the receipt of a case be confirmed within a certain timeframe and that the involved parties be regularly informed about progress. The Schlichtungsstelle Mobilität lists an average processing time on its website.

Efficiency also includes guaranteeing the board's ability to:

- gather the necessary *information* for judging the case;
- form a constructive *cooperation* with the company and the organizations involved;
- and remain in the position to fulfil its tasks even if the relationship to the company becomes strained, because differences of opinion do tend to arise.

In the cases examined, it was usually possible to develop a constructive cooperation. However, not all boards examined seem to be well equipped for the case that the company should fundamentally refuse to cooperate. The instrument of public pressure is indeed for the most part available, but in concrete cases it does not help much. Relatively often, cases are repeated if necessary and/or submitted to higher authorities, which can occur by way of a statute or informally. In contrast, a formal right to information – such as access to records, hearing of witnesses, or external advisory opinions – is not provided in every situation. In this regard, German boards are particularly badly equipped.

No conclusions about *legality* can be made in the context of this article but there is no sign that the arbitration services described here violate consumers' rights. It is a similar situation for *freedom of action* (binding force of arbitration settlements) and *representation* (calling in a third party or expert). In terms of freedom of action, it can be noted that hardly any of the examined boards can make decisions that are *binding for both sides* (Table 2.5).

2.5.10 Future Prospects

One of the fundamental aims of complaint management is to minimize the number of dissatisfied customers, but on the other hand to maximize the percentage of dissatisfied customers who do complain.[128] In an analogous manner, arbitration boards should publicize their service well, providing for a low-threshold inflow in order to contribute to customer satisfaction.

Following the understanding of alternative dispute resolution as a court of appeals, ADR bodies only deal with the tip of the iceberg of the most complicated or controversial complaints. It is thus a complement to other strategies for interest representation and consumer protection. Efforts for improved service and increasing customer satisfaction, however, have also to deal with the

128 See Stauss and Seidel (2002), p. 51.

Table 2.5 Status of arbitration results

Status	Board
Recommendation	Schlichtungsstelle Nahverkehr (G), LVB Ombudsperson (G), Schlichtungsstelle Mobilität (G), Bavarian Ombudsboard (G) Ombudsdienst SNCB (B), Vlaamse Ombudsdienst (B), Médiateur Région Wallonne (B), Bus Appeals Body (GB), BUCT Scotland (GB) ARN (S), Ombudsboard for PT (CH)
Recommendation, binding as a voluntary commitment of the company	Baden-Württemberg Ombudsboard (G)
Binding for the company as a negotiation outcome	Service de Médiation STIB Brussels (B), Médiation SNCF (F), Médiation RATP (F) Passenger Focus (GB), LondonTravelWatch (GB), General Consumers' Council Northern Ireland (GB)
Binding for the company	IPFAS (GB)
Binding for both sides	SGC Openbaar Vervoer (NL)

Source: author's depiction based on information provided by the ADR bodies.

question of how to solve these difficult disputes as amicably as possible. The company ultimately profits from this in the form of avoided trouble and avoided legal proceedings. In this respect, the ADR instrument offers a wide range of advantages (see Sections 2.5.1/2.5.2) which can be brought to bear in PT as well. This is shown in examples across Europe. But the aforementioned EU quality criteria need to be better observed in some countries so that arbitration can make an effective contribution to customer satisfaction.

2.6 Passenger Rights – Principles of Legislation in Europe

Mihael Pohar

2.6.1 Passengers' Rights in Germany

2.6.1.1 Overview In Germany, there are several types of legislation that regulate the contractual relationship between passenger and transport company. These include laws, statutory ordinances, and transport companies' general business terms and conditions (terms of carriage). Box 2.1 gives a very simplified summary.

2.6.1.2 Special regulations for public passenger transport In the legal foundations for public transport, a distinction must be made between rail transport and other forms of public transport. The Allgemeines Eisenbahngesetz (AEG) applies to rail transport. For other forms of local transport, including buses and street trams in regular service, the Personenbeförderungsgesetz applies (PBefG). The civil rights determined in Bürgerliches Gesetzbuch (BGB) apply to all transport contracts regardless of the form of transport; the policies in the BGB are, however, mostly comprised of the specific regulations in the AEG and the PBefG.

Box 2.1

Bürgerliches Gesetzbuch, Civil Code (BGB) – applies to all types of transport contracts
Allgemeines Eisenbahngesetz, Railway Act (AEG) – applies to rail transport within Germany
Personenbeförderungsgesetz, Passenger Carriage Act (PBefG) – applies to carriage in other forms of local transport within Germany
Convention Concerning International Carriage by Rail (COTIF) – applies to international rail transport
Uniform Rules Concerning Contracting for International Carriage of Passengers and Luggage by Rail (ER CIV) – applies to international rail transport
Eisenbahnverkehrsordnung, Rail Transport Ordinance (EVO) – special regulations for national rail carriage
Ordinance for General Terms of Carriage (Allgemeine Beförderungsbedingungen) (VO AllgBefBed) – special regulations for carriage in other forms of local transport within Germany
Transport Companies' Terms of Carriage – each individual company's specific contractual regulations

Allgemeines Eisenbahngesetz, Railway Act (AEG): the Allgemeines Eisenbahngesetz (AEG) of 27 December 1993 applies to railways. The AEG regulates the basic parameters for rail companies and determines, for example, that the operation of a public rail company requires authorization. It also stipulates that rail companies are fundamentally required to make carriage contracts with everyone. The AEG was essentially reformulated in 1993 with the rail reform in Germany and the conversion of Deutsche Bundesbahn (West German Federal Railway) into Deutsche Bahn AG (German National Railway, plc).

Eisenbahnverkehrsordnung, Rail Transport Ordinance (EVO): the Eisenbahnverkehrsordnung (EVO) regulates particularly the closer relationship between passenger and railway company. It fulfils the special statutory regulations provided by the AEG and was enacted by the Bundesverkehrsministerium (Federal Ministry of Transport) with approval/changes from the Bundesrat (Federal Council). The origin of the current EVO is, however, an Act from 1938. The first Eisenbahnverkehrsordnung (EVO) was established as early as 1908. A central point determining liability in cases of delay or cancellation in train service was a regulation that existed from this time until the year 2006, Section 17 EVO, which stated:

> Delay and cancellation do not constitute a claim for compensation. However, the railway must, as much as possible, provide for passengers' further transport in cases of cancellation and impeded continuation of a journey.

This complete exclusion from liability, which was a considerable advantage for rail companies, was modified on 1 July 2006. The starting point was a change in international rail law, namely the 1999 changes to the COTIF Convention. Since then Section 17 (1) has stated:

> The rail company is liable for damages that originate from train cancellation, delay, or missing of connections that make it impossible to continue the journey on the same day or, under the given circumstances, make the continuation of the journey on the same day unreasonable. Compensation includes reasonable costs that arise through overnight accommodation and notifying the parties awaiting the passenger.

Thus the passenger's new legal position now includes a claim to compensation for costs if a planned journey can no longer be completed on the same day. In cases of delay or cancellation in same-day transport there still exists, however, no legal claim to compensation. Experts are intensively discussing whether this kind of extensive exclusion of consumers' rights is compatible with the German Constitution.[129] In legal practice, however, there have to date been no recognizable trends toward deeming the EVO unconstitutional in any of its points.[130]

Public passenger transport based on the PBefG and VO AllgBefBed: the Personenbeförderungsgesetz (Passenger Carriage Act (PBefG)) applies to most forms of public passenger transport with route-based buses and street trams.[131] Based on this Act, the Ministry of Transport issues the 'requirements for the general terms of carriage for street trams and omnibus transport as

---

129 See in particular Staudinger (1999); Engel (2002); Faulenbach (2005); Däubler (2003); Bitter (2005), Pohar (2006); Schmidt-Bendun (2006).

130 See most recently OLG Karlsruhe RRa (2005), p. 280; see also LG Frankfurt a.M. NJW (2003) p. 3611; LG Essen RRa (2004), p. 32; AG Berlin-Lichtenberg TranspR (2001), p. 212; AG Frankfurt a.M. TranspR (2000), p. 313.

131 From 8August 1990 (BGBl. I, p. 1680), last changed by law on 19 July 2002 (BGBl. I, p. 2691).

well as route transport with motor vehicles' (VO AllgBefBed).[132] It is up to the transport companies whether they decide to supplement the requirements with additional regulations of their own in the form of so-called special terms of carriage.

Terms of carriage: with terms of carriage, rail companies or other transport companies shape the legal relationship between themselves and their passengers. The terms of carriage represent general business terms and conditions that are included in the ticket sales section of each respective transport contract, and are thus explicitly named in the contract itself. Using the terms of carriage, a passenger can see if he has a claim to compensation if a train is delayed, does not arrive, or if another type of travel disturbance occurs. In order to strengthen their customers' interests, many terms of carriage provide claims to compensation that would not exist based solely on legal foundations.

Deutsche Bahn's regulations, for example, are called *Beförderungsbedingungen Personenverkehr* (Terms of Carriage for Passenger Transport). As DB has done, other transport companies or federations have also introduced ex gratia payment regulations or voluntary self-requirements. For most companies or federations, these regulations have become part of their general terms of carriage. Under certain conditions, passengers receive compensation or the option of using another form of transport (for example, a taxi).

In 2002, Deutsche Bahn changed its terms of carriage in terms of liability in cases of cancellation and delay. Since then, passengers have had a claim to compensation for damages if a train does not arrive or is late. This regulation corresponded with the Section 17 EVO, which was passed four years later in July 2006. However, in its own terms of carriage, DB solidified the term 'on the same day' to 'until 24:00' and 'appropriate costs' were limited to a 'maximum amount of €80'.[133]

According to its terms of carriage, Deutsche Bahn additionally provides reimbursement of 20 per cent of the value of a ticket if a trip on an ICE or IC/EC train is delayed more than 60 minutes. This also applies if the delay was due to cancellation of a train within a network of connections. The amount of the reimbursement must be at least €5. The passenger does not, however, receive cash, but instead a coupon which is valid for a year and can be counted against the price of another ticket.[134]

International Rail Carriage – COTIF, CIV and TCV: important legal foundations in international rail transport include the COTIF Convention, the uniform rules of the CIV and the fare TCV. The COTIF Convention was

132 From 30 June 1989 (BGBl. I, p. 1273), last changed by lay 15 December 2001 (BGBl. I, p. 3765).

133 Section 9.1 BB Personenverkehr.

134 Section 9.1.4 et seq. BB Personenverkehr.

passed on 9 May 1980.[135] For the Federal Republic of Germany, the Convention came into effect on 1 May 1985. Particularly relevant for passenger transport are the uniform rules of the CIV.[136] These regulations are international law, which overrides national law. For rail journeys that cross a national border, alongside the uniform rules of the CIV, transport companies apply the Common Fare for the Transport of Passengers and Luggage (TCV) from 12 December 2005.[137] The international fare is also part of general business terms and conditions or terms of carriage within each country. The German Eisenbahnverkehrsordnung also expressly states in Section 1 that the common international fare overrides the regulations of the EVO. In other forms of international transport as well, international law overrides the regulations that normally apply within Germany.

2.6.1.3 Bürgerliches Gesetzbuch (Civil Code (BGB)) As special norms, the laws specifically addressing public transport, that is the AEG, EVO, PBefG, and VO AllgBefBed, override the regulations of the BGB. However, the BGB cannot be overlooked because a contract for carriage between a passenger and a company is a contract bound by civil law, and thus the regulations of the BGB take effect when a legal question is not already regulated by the special norms.

A contract for carriage is a contract for services in the sense of Sections 631 et seq. BGB. Most rules in contracts for services, such as a guarantee claim (for example a right to mitigation in cases of poor service) are regulated by the EVO and the VO AllBefBed. However, there are some legal questions that are ultimately answered by the BGB (see Box 2.2).

2.6.1.4 Passengers' rights in route-based bus and light rail transport in Germany Section 16 VOAllgBefBed applies to transport with buses and street trams in cases of delay of a vehicle. Based on this regulation, customers have no legal claim if the timetable is not upheld. Likewise, service disturbances or

135 The member countries at that time included: Albania, Algeria, Bosnia and Herzegovina, Belgium, Bulgaria, Denmark, Germany, Finnland, France, Greece, Iraq, Iran, Ireland, Italy, Croatia, Lebanon, Lichtenstein, Lithuania, Luxemburg, Morocco, Macedonia EJR, Monaco, The Netherlands, Norway, Austria, Poland, Portugal, Rumania, Sweden, Switzerland, Serbia, Slowakia, Slowenia, Spain, Syria, The Czech Republic, Tunesia, Turkey, Ukrania, Hungary, United Kingdom; see www.otif.org.

136 OTIF agreement.

137 The abbreviation TCV stands for the French name 'Tarif commun international pour le transport des voyageurs et des bagages'. Printed in # 701 of the Fare Directory, see also www.bahn.de.

Box 2.2

- Issues concerning closing of contracts (right to representation, minors)
- Requirement for information, liability for situations resulting from deficient information (for example in cases of delay of a vehicle)
- right to information about pre-existing conditions
- A passenger must be comprehensively and completely informed about various connections and travel fares before beginning a journey.* If this does not occur, the company may be liable for compensation of damages.
- A passenger has a claim against a transport company if he has reserved a seat and this reservation was not kept by the fault of the company.
- A company can have a right to compensation for damages from a passenger as well. For example, if a passenger brings luggage that is not permitted to be transported and this luggage causes damages.
- The company is liable for breeches of so-called contractual accessory obligations. Such a contractual breech exists, for example, if the toilets** in a train are unusable. It is conceivable that a deficiency in climate control on a particularly hot summer day could also result in such a liability. This liability must, however, be determined on a case-by-case basis.
- The contractual relationship between transport operator and infrastructure provider is also regulated by the BGB (Section 278). The infrastructure company is hence considered an auxiliary of the service provider.

* For more information see Pohar (2006).
** AG Frankfurt a.M. NJW 2002, 2253 f.

disruptions as well as lack of seating do not constitute a claim for compensation. Companies are also not liable for maintaining connections.

However, numerous companies have their own ex gratia payment regulations (see Section 2.4). The Transport Association of Rhine-Ruhr (VRR), for example, offers a mobility guarantee. According to this guarantee, taxi costs are reimbursed if a vehicle departs with a delay of more than 20 minutes and there is no replacement connection. Similarly, in the context of a mobility guarantee, the Transport Association of Rhine-Sieg (VRS) offers to cover taxi or even ICE costs in cases of more than 20 minutes of delay. The BVG in Berlin also offers passengers a free ticket in cases of more than 20 minutes of delay during the day and covers taxi costs of up to €25 during the night.

2.6.2 Passengers' Rights in Austria

In Austria, according to the Eisenbahnbeförderungsgesetz (Rail Carriage Act (EBG)) of 10 March 1988, there is a legal exclusion from liability that is similar to the German Section 17 EVO. It is expressly stated that there is no legal claim in cases of train cancellation or delay, if a train departs with a delay, arrives with a delay, or if a train does not arrive at part of a route, or at all (Section 24 EBG). Only if a connection is missed is there any legal claim at all.

According to the Kraftfahrliniengesetz, Road Services Law,[138] which applies to route-based buses, claims for compensation in cases of cancellation or delay are excluded. Companies are, however, allowed to provide their own additional special terms of carriage, which must be submitted to the appropriate office for authorization. For street trams, underground trains, and omnibuses, there is a separate body of laws. They are subject to the Eisenbahngesetz (Rail Act) of 1957. According to this law, every transport company is required to create their own fare, which is to be officially authorized if it does not conflict with public interests. Companies are generally excluded from liability for delay or cancellation of street trams in Austria's terms of carriage:[139]

> The failure of a vehicle to depart or make a connection, delays in departure or arrival, as well as service disturbances, service disruptions of any kind, and lack of seating do not constitute a claim for compensation for damages as long as the damages were not caused by deliberate or grossly negligent behaviour on the part of the transport company's employees. If the passenger is not entitled to a claim for compensation for damages in a particular case, there is to be neither a reimbursement of the travel fare or transport free of charge. In cases of delay and cancellation of vehicles, as well as lack of seating, however, the transport company is required to transport the passenger with the next suitable vehicle on the same or another route without additional charge such that the passenger can reach his destination with the least possible delay.

2.6.3 Passengers' Rights in Belgium

In Belgium, there are generally no general terms and conditions concerning the relationship between a passenger and a rail company; instead, the Belgian federal railway, SNCB, imposes relatively strict requirements on itself, the *conditions générales*. In the course of conforming to the new conventions for international rail transport (COTIF 1999 and ER CIV 1999), in Art. 12 Section 2 of the *conditions générales* a requirement was added which states that the SNCB must either bring the passenger to their destination with a taxi or pay for accommodation in a hotel if the last train cannot be reached due to a delayed connection. The SNCB decides for itself which form of

138 Federal law concerning route transport of passengers with motor vehicles (Kraftfahrliniengesetz – KflG).

139 See www.vor.at.

compensation they offer to the customer. The regulation accords primarily with Deutsche Bahn AG's terms of carriage for passenger transport. One special aspect of mention is that in Art. 58 Section 2 in addition to a claim for replacement transport or overnight accommodation, the passenger is additionally guaranteed an extensive claim to compensation for damages in cases of train delay, cancellation, or missed connections. This guarantee is limited to the cost of the ticket and is provided if the passenger can provide proof of damages that are due to the fault of the transport company.

2.6.4 Passengers' Rights in The Netherlands

In The Netherlands, the Dutch Civil Code (Burgerlijk Wetboek) regulates the liability of transport companies. Transport companies can further shape their legal relationship through general terms of carriage (Algemene voorwaarden). In this way, the Nederlandse Spoorwegen (NS), for example, is excluded from liability in cases of train cancellation or delay. In the terms of carriage for transport with common public vehicles, the Burgerlijk Wetboek also excludes liability for train cancellation and delay. NS does, however, have generous ex gratia payment regulations. In cases of delay of more than 30 minutes, half of the travel fare is reimbursed, and if a train is more than 60 minutes late, the entire fare is reimbursed. Of particular mention is the fact that NS transfers the reimbursement directly to the customer's account and does not issue a circuitous coupon like Deutsche Bahn.

2.6.5 Discussions of Passengers' Rights – Proposals for Reform

In discussing passengers' rights, a distinction must be made between plans on the European level and national legislative efforts. At the European level, Ordinance 1371/2007, passed on 3 December 2007, has been announced in the Official Journal. It will come into effect two years after its announcement. In Germany, how passengers' legal position should be improved is being discussed at the federal and country levels.

2.6.5.1 The discussion of reform in Germany Because liability for train delay and cancellation was completely excluded in Germany according to Section 17 EVO up until 30 June 2006, passenger associations and consumers' rights organizations[140] have long been calling for passengers' rights in national passenger rail transport to be strengthened legally.[141] Many customers have found the privileges in liability for rail companies to be out-of-date. Since 2002, the issue of passengers' rights has been on the political agenda on the federal and country levels. The

140 See the publications of the Schlichtungsstelle Nahverkehr, Arbitration Board for Local Transport, North-Rhine Westphalia.

141 See the Pro Bahn Newsletter *derFahrgast*.

federal government has developed a 'quality push for public transport'[142] in which it fundamentally argues for improvement of passengers' rights.[143] Experts on the federal[144] and state[145] levels are discussing the rights of passengers in local and long-distance transport. The federal government has commissioned a report and published its results.[146] This report includes concrete proposals for legislative measures. More and more possibilities are emerging. It has been possible to change the EVO and to add special regulations for passengers' rights to the BGB. An independent law for passengers' rights is conceivable. Proposals to date have, however, been postponed pending developments on the EU level. A bill was announced on 7 September 2007 by the leading Federal Ministry of Justice.[147]

2.6.5.2 European Regulation Concerning Passengers' Rights and Requirements in Rail Transport The origin of Regulation 1371/2007 was the regulation proposal 143 from the commission KOM (2004) from 3 March 2004, which was presented in the context of the third railway package.[148] The regulation is meant, among other things, to serve quality and efficiency in rail transport service, particularly in order increase the transport market share of railways in comparison to other transport providers.

142 BT-Drs. 14/9671.

143 See the 'Qualitätsoffensive im öffentlichen Personenverkehr' ('Quality Push in Public Transport'), BT-Drs. 14/9671, as well as the petition of the *CDU/CSU-Fraktion* from 23 April 2002, BT-Drs. 14/8853; or the minor inquiry of the *CDU/CSU-Fraktion* from 3 July 2001, BT-Drs. 14/6610.

144 Such as the hearing 'Die verbraucherpolitischen Aspekte im öffentlichen Personenverkehr' ('The Transport Political Aspects in Public Transport') by the Committee for Consumer Protection, Nutrition, and Agriculture in Berlin from 5 May 2003.

145 Such as the congress 'Verbraucherrechte im öffentlichen Personenverkehr; König Kunde bei Bus und Bahn?' ('Consumers' Rights in Public Transport; Is the Customer King on the Bus and Train?') by the Ministry for Environmental and Nature Conservation, Agriculture, and Consuemr Protection of the state North Rhine-Westphalia from 12–13 June 2003 in Bonn.

146 See BT-Drs. 16/1484.

147 http://www.bmj.bund.de/fahrgastrechte, accessed November 2007. The bill was not available before this book went to print.

148 See the announcements of the *Commission* KOM (2004) 140; KOM (2004) 139; KOM (2004) 142, KOM (2004) 143; SEC (2004) 236, all from 3 March 2004; available at: http://europa.eu.int/comm/transport/rail/package2003/new_en.htm.

This is intended to continue the 'integration of the European rail system',[149] which began with the first[150] and second[151] rail packages. Regulation 1371/2007 also follows the 'consumer political strategy 2002–2006'[152] in creating a high consumer protection level in transport. The regulation proposal was preceded by the Commission's White Paper 'European Transport Politics until 2010 – Directions for the Future'.[153] This was the first time that the 'support of measures to guarantee the quality of rail transport services and ensuring of passengers' rights'[154] were described as a concrete goal for European Community policy. On 4 October 2002, the Commission presented a consultative paper: 'Passengers' Rights and Obligations in International Rail Transport'. With this paper, the Commission called for consumer organizations, unions, and rail transport providers to make statements. While passenger representatives welcomed the push, particularly transport providers and infrastructure companies, including Deutsche Bahn AG,[155] looked at the proposal critically and remained distant from it.[156] The political groups involved included this attitude in the commission proposal. While passenger associations approve of the proposal and say that it is not extensive enough, providers object to it.

EU Regulation 1371/2007 comes into effect on 3 December 2009 directly as national law. Until then, national lawmakers are required to act in accordance with existing national legal foundations.

2.7 Conclusion: Taking the Customer Seriously

Martin Schiefelbusch

As a general rule, only a small part of a company's customer contact leads to complaints or to any kind of legal dispute. In this respect, it may be assumed that it is irrelevant to pay so much attention to such cases. Indeed, this assumption does not look far enough: only a small percentage of dissatisfied customers take it upon themselves to complain, and the effect of a negative (service) experience is exponentially greater than a satisfactory service, which is simply expected, without being noted as positive (see Section 2.2.2). Bad experiences do not

149 See the announcement of the Comission 'Continuation of the Integration of the European Rail System – Third Rail Package'; KOM (2004) 140 from 3 March 2004.

150 Principles 2001/12/EG; 2001/13/EG; 2001/14/EG.

151 See. KOM (2002) 21; KOM (2002) 22; KOM (2002) 23, all from 23 January 2002.

152 KOM (2002) 208; ABl. C 137 from 8 June 2002, p. 2.

153 From 12 September 2001, KOM (2001) 370.

154 KOM (2001) 370, p. 125.

155 Statement of DB AG from 20 December 2002.

156 KOM (2004) 143, p. 4.

simply remain in the memories of those affected, they are also passed along – to one's circle of friends as well as to the general public. Bad news is often more interesting to the media than, perhaps, the fact that 90 per cent of customers are satisfied or that a new quality level has been successfully reached.

These basic rules apply to public transport, as well as to other industries. Why then is transport comparatively so poorly rated in inter-industry measurements of satisfaction? The reasons for this fact can be sought on several different levels. First, in comparison to other fields and in the view of passengers, transport companies' customer service leaves room for improvement. The issue of passengers' rights is one among many aspects, but it is an essential part of the formal foundation on which company decisions build the positioning of their customer service.

Travelling with PT is also significantly more defined by other people than individual mobility: timetables, which are always a compromise between a multitude of needs, must be observed. During the trip, the travellers give themselves over to the 'system' and are obliged to be largely passive.[157] The common trip with other people (often strangers) contributes to the influences of others on the experience of the service. Thus public transport services offer their users significantly fewer possibilities for emotional as well as tangible personalization than individual transport, which offers these possibilities as fundamental to its appeal. As long as psycho-social needs such as image promotion, self-expression and privacy regulation[158] are not met by the transport service, there is an even greater predisposition to be critical of the service.[159] Identification with and emotional connection to the service must thus be developed in other ways. An important contribution to this is provided by participatory opportunities in service planning, as will be presented in Chapter 4.

It is also important to apply the standard 'We take our customers seriously' in service provision. The aforementioned passivity in PT is interpreted differently according to how agreeable the journey is. If waiting is, in this regard, already perceived as a downside, the feeling of disadvantage is even stronger if the passenger has to wait on a crowded pedestrian island while car traffic flows around them unhindered.[160]

The term 'dignity' has been absent from the aforementioned psycho-social requirements of mobility services. It may seem abstract, but it is a fundament of satisfaction and a binding element between the various levels addressed here. Ultimately, it is always about the customer's *sense of value*.

There are many ways of dealing with this issue to be found in this chapter. The space of this book does not allow an examination of every transport

157 Bliersbach (1992), pp. 32–3: cf. Karopka et al (2000), pp. 20–21; Meier-Dallach (2004).

158 See Klühspies (2001).

159 See infas/DIW (2004), p. 100.

160 See Hilgers (1994); Klühspies (2001).

company's widely differing positions on this topic. As is often the case, the presentation here of extreme positions is not true to all the facets of reality, but is necessary due to spatial constraints. Likewise, we have put some issues aside that have not been fully realized in contemporary practice but nevertheless seem important for the question of value and satisfaction:

- *Consequences of complaints.* Customer criticism offers a company the possibility to offer redress after a dissatisfactory experience; it also offers an opportunity for learning (see Section 2.2.1). Whether this actually happens depends not only on effective complaint management (see Section 2.2.3), but also on how the company values the information accumulated in this manner. This particularly concerns the specialist departments (planning, operations and so on) whose orientation on fundamental long-term issues does not necessarily accord with the experiences of the everyday user.[161]
- *Quality versus speed.* A quick response to complaints is often seen as the central quality criterion, which is why direct complaint management – essentially independent of the company's size – is to various degrees standardized, and in some cases even automated. The limitations in dealing with the complaint's content in this priority structure (see Section 2.3) seem less than fully considered. A reply that only partially addresses the complaint's content, contains a justification of the occurrence in technical terms, and perhaps even contains grammatical errors, does not create the impression that the company has dealt with the concern, even if it reaches the customer within a couple of days. In this regard, small and medium-sized businesses have potential advantages if they take advantage of their closeness to the customer and their better communication with specialist departments in creating more personal replies.
- *Addressing other quality criteria.* The creation of service guarantees (see Section 2.4.4.1) as well as the discussion of passengers' rights both concentrate on delays, cancellation, and partially on missed connections. Punctuality is without doubt a central quality criterion in transport services. In light of the criticism that is common in other areas, it should be considered whether similar quality-securing concepts could be developed for passenger information and comfort.
- *Immediate aid versus retroactive compensation.* For obvious reasons, problems seem to be greatest at the moment of their occurrence – and not just because a fraction of affected customers ultimately complain (see above). In all (legitimate) activities in complaint management and passengers' rights it should thus be considered how to provide aid or at least relief as soon after the problem as possible. This addresses

161 See Chapter 4; Rajé (2007).

operations and quality management (see Chapter 3), but the formulation of guarantees and similar measures provide the opportunity as well to directly satisfy customers who experience problems (such as in a connections guarantee).[162]

In the field of customer service, as well as in its integration with service planning and service provision, there are thus numerous possibilities not only to employ instruments such as guarantees and arbitration services more commonly, but to further develop them in view of the challenges facing public transport.

162 See Section 2.4.4.1; Neugebauer (2007).

Chapter 3
Customer Perspective in Quality Management

3.1 Quality Management in Public Transport

Hartwig Meier

Passengers value punctuality, safety and cleanliness in buses and trains. This seems self-evident. However, the possibility that there are various interpretations for these terms implies problems: even 'punctuality' becomes a question of definition, but it is at least clearly quantifiable. What, on the other hand, is an objectively clean bus or a friendly, well-informed driver? The quality management teams in public transport and particularly in local public transport (LPT) ask themselves these questions.[1]

Two recent lines of development have markedly increased the relevance of quality management in public transport. Firstly, the service sector has been generally introducing quality management methods and standards; secondly, the transport market in Europe at large is undergoing severe structural changes.

With 'deregulation' as a key word, the second point especially has provided increased momentum for transport authorities and transport companies, as well as passenger associations and unions to concern themselves with aspects of service quality. It can safely be assumed that the increasing importance of competition has sharpened the awareness that not only the price but also the quality of a service determine the customers' acceptance of it. How ready the customers are to pay the price that accompanies, this is a different story.

According to careful observation, it is not justifiable to assume an equation wherein a higher degree of competition automatically brings about an increase of quality in all fields. In European transport, in both insulated, secure markets and deregulated markets, we find all levels of quality. In both cases, however, we find systems with customer-acknowledged value as well as providers with drastically limited basic services which are only used by customers who turn to public transport because they have no alternative.

While not desiring at this point to go into the inexhaustible discussion of legal foundations and of LPT's organizational structures, it should be

1 According to the overall theme, many of the statements apply to local public transport as well as long-distance public transport. LPT is the central subject of the following remarks.

emphasized that it is not competition, but the common (if not, however, inevitable) moment when the *definition of quality standards* becomes consciously incorporated into this discussion that represents leverage for improvement of customer orientation. Independent of transport contracts allocated by competition, appointment, or direct contract awards, the public increasingly expects fixed standards and is assured of receiving them through transport companies that work in a customer-oriented way.

At the core, not only are the passengers' expectations revealed through the customer perspective in quality management, but above all, the transport company and transport authorities are forced to make a conscious *decision about the quality levels* they are striving for – they are forced to acknowledge customer expectations. These decisions define what will go onto the streets and onto the tracks as services. For this reason, Section 3.4 will make a thorough exploration of the importance of measuring customer satisfaction.

A further specific aspect of public mobility presents the fact that a high number of LPT users also have access to individual automobility. For this reason, a *constant comparison* with the private car underlies the quality of services in personal transport. The car serves customers in view of numerous quality criteria (comfort, accessibility, availability etc.) as a reference point. This is a particular cause for significant increases in their expectations in recent years.

In contrast to a personal car, to which not only aspects of mobility but lasting product features can be allocated (including status), service in LPT is perceived as a *snapshot* in which it is rarely possible to compensate for dissatisfaction.

The following sections are meant to portray what the terms 'quality', 'quality management' and 'quality management system' stand for, how they are specifically understood in the public transport service and what significance customer perspective has there. After a short overview of the genesis of this kind of quality management thinking (see Section 3.2), the core of the newest developments in European standards will be presented with the standard family of the EN 13816[2] (see Section 3.3). Their basic methods will be sketched out and the importance of customer satisfaction measurement in quality management will be demonstrated (see Section 3.4). Finally, examples for the role of passengers in quality management will be presented (see Section 3.5).

2 Complete: EN 13816. In the following, the common abbreviations are predominantly used for standards in the context of this text: the standards that apply to Europe are referred to as EN and those that apply worldwide as ISO. The abbreviation is usually preceded by a country-specific code like DIN in Germany, AFNOR in France and BS in the UK.

3.2 Fundamental Terms in Quality Management and their Meaning in LPT

Hartwig Meier

A look at the essential definitions in quality management and an overview of the genesis of quality management thinking will help to classify and understand the resources currently available for quality management in LPT.

The term *quality* can be translated from the Latin source (*qualis*) by 'as constituted'. In general, quality describes the constitution, the 'goodness' or the worth of an object. Because of the highly versatile facets of the term quality, the definitions encompassing all areas of application are accordingly cumbersome:

The fundamental standard for quality management, ISO 9000, describes quality as the degree to which a unit fulfils the demands of its inherent criteria. Quality management serves coordinated activities towards leading and guiding an organization in terms of this quality, and a quality management system is ultimately a management system for leading and guiding an organization in terms of quality itself.[3]

As used in LPT, quality is understood as the merit in terms of various criteria of service quality. The Verband deutscher Verkehrsunternehmen (VDV) defines it, for example, as 'the degree of concordance between the performance criteria of a product or a service and the according performance expectations of the customers which the respective company would like to reach as a target group'.[4]

In more lay terms, quality describes simply the suitability for usage. In order to achieve this, the principles of orientation to the customer, target-oriented leadership and the inclusion of the people who work for the service all must be considered. A *quality management system* works to present to an organization how it can consistently provide services and thereby meet the customers' demands as well as legal conditions. The necessary steps toward effective enhancement of customer satisfaction must be established.[5]

The development of this system, which is currently in wide use, began approximately 50 years ago. Quality management systems originally were conceived for workshops, thus the first national and international standards were concentrated on avoiding errors and accordingly on recognition and non-shipment of faulty products. However, this idea should not be carried over to the service sector: a faulty product can be recalled but a faulty service has already reached the customer the moment it has occurred.

Alongside other changes not presented here, the customer perspective has been essentially better observed in the newly modelled EN 9000 series since 2000. One specific problem in adopting these standards in this service sector,

3 EN ISO 9000 (2005), Quality Management Systems – Foundations and Terms.

4 Association of German Transport Companies (1996).

5 EN ISO 9001 (2000), Quality Management Systems – Requirements.

however, remains – the problem of the ambiguous 'double customer' in LPT: the passenger as well as the transport authorities can be the customer in the sense of the ISO 9000.

This problem, as well as the desires expressed at the European level to be able to provide more concrete criteria for LPT, led to the formulation of EN 13816. Specialists from transport companies, transport authorities and their representative associations from various European countries collaborated on the design until 2002. While ISO 9001 provides the conditions for integrated quality management systems for all companies, the *European Quality Standard* describes a guideline for product-specific quality management through concrete recommendations towards sensible quality criteria and measurement procedures for entities in public transport (local as well as long-distance transport), and in this way aims to contribute to stabilization and improvement of the quality of LPT.

The customer-orientation supported in ISO 9001 is underlined in EN 13816 through the introduction of a service quality loop, which will be more specifically dealt with in the following chapter. Contrary to ISO 9001, EN 13816 is not organization- or company-oriented, but product- and service-oriented.

Based on EN 13816, in 2006 EN 15140 appeared as a supplement that addressed the measurement of the quality of services delivered and, additionally, provided helpful examples.

In the overview of the jungle of standards, the following rule is helpful: for forming a quality management system in a company or an organization, the orientation provided by the ISO 9000 series is helpful; for a perspective on customers and LPT content specifications, the 'family' of EN 13816 is the suitable instrument.

Who, though, is the *customer in local public transport*? Complete coverage of the costs of LPT from travel fares with regard to the requirements of transport, social and environmental politics is currently rarely possible. Federal governments, states, and municipalities contribute in various ways to the financing of LPT. This leaves room to suppose that alongside the 'passenger' customer, the actual users of the service to be observed as customers of the transport company are these bodies.

Because the allocation of these financial means is always aligned with the suitable utilization by the current or desired customers, in the sense of quality management the term 'customer' is to be understood solely as the *passenger*. A transport company's service is always ultimately justified by the passengers' utilization of it. The transport authorities have to make their efforts towards quality in the organization of public transport passenger-oriented, in that they too are service providers for the customers (whether it be through direct announcements or indirect long-term local transit plans).

The transport authorities' quality management systems distinguish themselves from those of transport companies most notably in their choice of criteria and (for the most part) more limited differentiation. In each case, it makes sense to gather quality data in areas where it is possible to improve customer satisfaction. The fact that the transport authorities consult these data as verification of the fulfilment of transport contracts has a synergetic effect. Without an answer to the question, 'Who can make sure specific deficiencies do not recur?', it is not a matter of quality management systems, but merely of accounting.

If at times an image arises of a very complex arrangement between transport companies, infrastructure operators, and transport authorities, then the orientation of the quality management system – at least in the eyes of the customer, who has no interest in organizational subtleties – can be reduced in complexity. Ultimately the question remains: 'What does the customer really want?'

3.3 Service Quality in Public Transport: EN 13816

Hartwig Meier

The European standard for definition and regulation of the objectives and the measurement of service quality in public transport[6] expresses its intention as follows:

> The main purpose of the standard is to promote a quality approach to public transport operations and focus interest on costumers' needs and expectations ....

The processes to be regulated for this aim should guide attention to essential points, consolidate decisions, enable quality comparisons and contribute to continual improvement. The implementation should '... allocate the resources available to the tasks most likely to produce added customer satisfaction and revenue to the service providers'.[7]

EN 13816 is (perhaps contrary to the general understanding of standards) *not* compulsory; it explains that it the standard 'may be appropriate for' reaching the aforementioned goals. This highly cautious approach is the result of a multilayered sphere of interest in its formulation. The standard subdivides itself into conceptual and methodical definitions and finally in requirements and recommendations (see Figure 3.1). The requirements are only binding if the standard is professed to be a valid standard by the operators.

6 EN 13816 (2002), Transportation – Logistics and Supply of Services – Public Passenger Transport, Service Quality Definition, Targeting and Measurement. All further references in this section refer, if not otherwise noted, to this standard.

7 EN 13816 (2002): both quotations are from the very extensive and programmatic preface, p. 4. The standard uses the abbreviation PPT for public passenger transport.

1. Scope
2. Terms and Definitions
3. Methodology
   3.1 The Quality Loop
   3.2 Quality Criteria
4. Requirements for Service Quality
   4.1 Compliance
   4.2 Quality Management
   4.3 Service Quality Definition
   4.4 Service Quality Measurement
5. Recommendations
   5.1 Commitment between Participating Parties
   5.2 Allocation of Responsibilities
   5.3 Other Quality Management Tools

Appendix A (informative)	Quality Criteria
Appendix B (informative)	Glossary of Terms pertinent to Public Passenger Transport
Appendix C (informative)	Guidance Notes on Performance and Satisfaction Measurement

Figure 3.1 Contents overview EN 13816

Source: DIN EN 13816:2002.

At heart this is already addressing a standard family – in addition to EN 13816, EN 15140 already presents a daughter standard that describes implementation regulations and specification of measurement systems. It is expected that in the future changes in the general conditions in the European LPT market will cause the pressure for unified regulation to increase and that theses special standards for service quality in public transport will be further differentiated.

The fundamental methodical thought process of EN 13816 is described visually in the quality loop (Figure 3.2). It subdivides service quality into four different perceptions:

1. *service quality sought* describes the customer's explicit or implicit expectations;

2. *service quality targeted* presents the level of quality to be stipulated by the service provider: it is determined by customer expectations as well as by inherent necessities and financial and technical limitations;
3. *service quality delivered* defines the quality achieved by the service provider. The measurement is performed by the service provider, but refers to the customer's perspective;
4. *service quality perceived* describes the customer's personal experiences with the service.

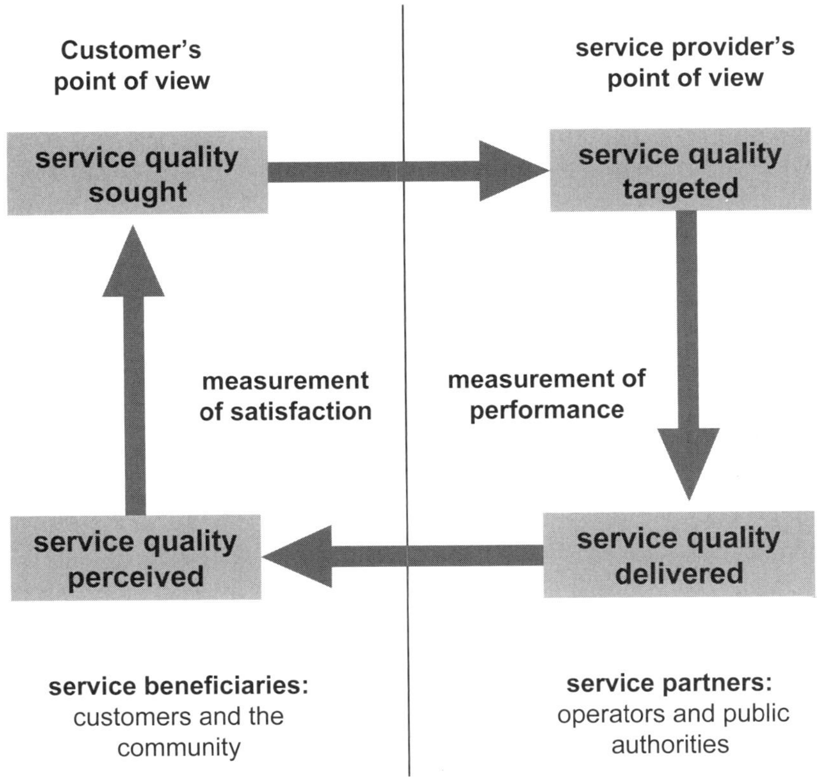

Figure 3.2 The quality loop according to EN 13816

Source: author's depiction.

The perspectives and the originators of the respective definitions are clearly named in both parts of the quality loop: what is important is that the passenger alone is clearly identified as the customer, while the transport authorities act together with the transport company as a service provider. This highlights the meaning of the responsibilities mentioned many times in the following standard text: for example, the transport authorities can have a massive influence on punctuality through the structure of bus or tram routes as well as traffic light priority measures. In the case of cancellation, the customer is rarely able to find the proper party to address (see Section 1.2.2). He or she can only evaluate the service provider complex in general. Who a passenger finds there depends on the point of entry: this should always be observed in the evaluation of transport companies and governmental organizations.

This qualification of the relationship between the public transport authorities (for example administrative unions, transport associations) and the transport companies as a customer-supplier relationship is, in the sense of this standard, not the subject. This, of course, does not put into question the fact that it does make sense to have quality criteria and to use quality management systems for relationships of this kind.

The quality loop describes not only the individual qualities, but also sheds light on the connections between them (Figure 3.3):

- the *difference between sought (or expected) and targeted (or intended) service quality* provides information about customer-orientation;
- the *difference between targeted and delivered service quality* is the measure for entrepreneurial performance (provided that the intended values are realistic);
- the *difference between delivered and perceived service quality* is an indicator of the importance of personal experiences, opinions and perhaps also prejudices;
- the *difference between perceived and sought service quality* presents customer satisfaction.

Implementation, too, follows a loop structure: at the beginning there is the determination of a usable and deliverable service with the according definitions of unacceptable performance (*threshold of non-acceptance*). At the core is the delivery of these services, the entrepreneurial measurement of them and finally the measurement of customer satisfaction. From both measurements, suitable corrective measures and analyses are presented which can once again change the definition of the services set forth at the beginning. If applicable, notifications of the results and developments are delivered to the customer.

The second methodical determination in the EN 13186 addresses the *quality criteria*. The quality of public transport is distinguished by several aspects. The standard organizes all possible criteria into eight categories:

1. availability;
2. accessability;
3. information;
4. time;
5. customer care;

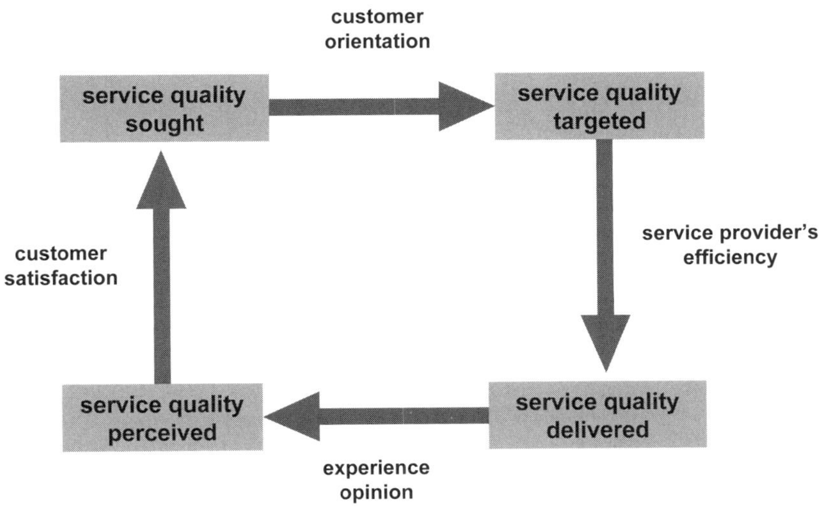

Figure 3.3 Correlations in the quality loop according to EN 13816

Source: author's depiction.

6. comfort;
7. security;
8. environmental impact.

The first two categories, availability and accessibility, describe the offering itself, while the next five present the quality of the service in detail. The last category assimilates the effects on the community at large. In the standard's appendix, examples are presented for further sub-categorization of these categories.

It is a sign of the standard's flexibility that it does not set any concrete parameters on individual criteria or assemble any checklists, but instead makes the involved parties responsible for these things. There are also no concrete quality levels named. This is very welcome in light of the high differentiation present in the public transport situations of the countries of Europe, the differences between local public transport and regional rail transport, and especially the differences between metropolitan and rural areas. This, however, places higher demands on the responsible parties on-site to conscientiously formulate their actions and leads to the danger of an arbitrary copying of transferable examples.

At the centre are structures, not concrete subject matter. Namely, the importance of local situations for the shaping of quality management in LPT is not to be undervalued. Criteria addressing passenger security, for example, will be of much more concern in major cities, for instance at tunnel stations, than in thinly populated regions; while in those areas criteria dealing with travelling comfort can be afforded above-average importance because of longer distances to commute. What the relevant criteria are and which target level is adequate can be significantly elucidated, as will be shown in the following section.

In the demands and recommendations of EN 13816, the described methodical approaches are translated into *operational guidelines* for quality management. In this, the following point is once more elucidated: for functional quality management, the recognition and incorporation of responsibilities is essential. Particularly in the railway field, it is very apparent that, without the network operator, transport authorities or rail companies cannot implement many measures for quality improvement alone.

Finally, self-critical inspections are always important to remember in the formation of quality management systems. In this line of work, there is a great temptation to formulate unnecessary codes and collect enormous data volumes whose worth is ultimately lost. Insufficient trust between service partners increases the danger that these systems, because of their potentially very complex structure, will break off and create new substantial cost pools in local public transportation. The purpose of quality management is never to be lost sight of: it serves to assure customer satisfaction, but not to create quality bureaucracies.

3.4 Quality Management and Satisfaction Measurement Practices

Hartwig Meier and Nancy Neugebauer

In EN 13816 three practices are emphasized which are of particular importance to the measurement of service quality:

1. mystery shopping surveys (MSS);
2. direct performance measurements (DPM);
3. customer satisfaction surveys (CSS).

The contents of these practices will be presented here in an abbreviated form. As long as the practices serve toward increasing delivered service quality, EN 15104 can be referred to as a complement to EN 13816, which specifies several descriptive examples for requirements and recommendation for measurement systems.[8]

The *mystery shopping survey* serves the appraisal of *delivered service quality* on the *right side of the quality loop*. Depending on the responsibility for specific quality criteria or on an application of results for the configuration and accounting of transport authorities, the transport companies as well as

8 EN 15140 (2006), Local Public Transport – Fundamental Requirements and Recommendations for Systems and Measurements of Delivered Service Quality.

the transport authorities can be responsible for this measurement practice. A third party can also be incorporated for its implementation.

This measurement practice is conducted especially for quality criteria that are not technically verifiable, for example evaluation of the cleanliness and condition of vehicles and stations or for evaluation of competence or conduct of service and transport personnel.

For the measurement of these quality criteria, various kinds of mystery shopping surveys come into question:

- checklists/questionnaires;
- interviews;
- test purchases.

Which practices are suitable and economical is to be determined by the parties responsible. The frequency of the measurement practices is also to be arranged accordingly. It is decisive that statistically valid data are available at the end. Adequate time should be planned for the concrete development of measurement practices because final versions of checklists, questionnaires and evaluation sheets only become available after a series of test runs according to experience. (An example of the implementation of quality tests will be examined in Section 3.5.2.2.)

As in the mystery shopping survey, *delivered quality* is also identified on the *right side of the quality loop* with the help of direct *performance measurements*. These internal measurements are for the most part performed by the transport companies, but can also be prompted by the transport authority or coordinated in terms of the type and comprehensiveness of the measurements.

Above all, these measurement practices comprise the quality criteria that are verifiable either through technical means or operating staff. The first in line here is punctuality, but guarantee of connections, availability of suitable vehicles,[9] and all criteria having to do with equipment and facilities belong in this category as well.

The measurement intervals can be highly varied according to requirements. Punctuality and dependability of connections should be documented most often in order to ensure a valid database. With the help of modern data collection, even a permanent collection is possible that no longer needs to make use of to random samplings. In cases of mere verification of equipment

9 This criterion is a good example of how the measurement practices mesh with one another. In direct performance measurement, one can present how often the scheduled vehicles could be provided. With the mystery shopper practice, the relationship between the demand and the seating offering can be ascertained. Measurement of customer satisfaction, finally, reflects these results in terms of the customer's perceptions and expectations.

and facilities (example: how many vehicles are available for one ramp?), an annual inquiry is quite sufficient.

Measurements of customer satisfaction refer to the *left side of the quality loop*. With these methods all quality criteria are measured lend themselves to evaluation by the passenger.[10] The results of customer satisfaction analyses influence the target values being defined and, if necessary, non-acceptance thresholds. In order to provide that the results of all stakeholders are recognized without restrictions, it is recommendable to carry out customer satisfaction analyses through a third party, for example market research institutes. They should be repeated every two to four years in order to continuously determine changes in customer desires and requirements.

In measuring customer satisfaction, the task is exclusively to determine perceived and expected quality from the passenger's view. In order to better involve the customer in the design of the inquiries, it is recommendable to carry out customer workshops (see Chapter 4). This allows the relevance of the inquiry criteria as well as the formulation of the questions to be tested as well. Pilot surveys pursue a similar purpose in initiating customer inquiries.

A distinction should be made between object-oriented and incident-oriented inquiry. In cases of *object-oriented* practices a judgement of satisfaction is composed which is determined by individual quality characteristics such as cleanliness, punctuality and friendliness. With the help of rating scales, the strengths and weaknesses of the company are then identified. An *incident-oriented* practice, on the other hand, refers to moments (incidents) in which the customer comes into immediate contact with the service provider. The passenger is questioned in open, personal interviews about context-related episodes and stories such as the experience of buying a ticket or the trip itself. Through these detailed analyses, concrete instances and causes for service deficiencies in individual phases of the service-provision process are determined. The operating costs of inquiry and evaluation are, however, accordingly high. Object-oriented inquiries, conversely, are significantly easier to administer due to the possibility of standardization. They also offer the advantage of applying time series to the individual quality characteristics and showing before-after comparisons.[11]

10 For example, environmental aspects often elude customer evaluation in measurement of satisfaction: even if the customer has the expectation that vehicles are equipped with modern emission control, hardly any customers will perceive the differences involved in catalysts and filtration systems and accordingly be able to evaluate them. Notwithstanding, inquiries can be made as to the weight of this criterion, and possibly customer expectations.

11 See application areas as well as advantages and disadvantages of the respective methods of customer inquiry as presented extensively in: Blümel (2004), pp. 66 et seq.; Pfeiffer (2001), pp. 48 et seq.; Probst et al. (2001), pp. 8 et seq.

EN 13816 expects of the service provider's quality management that it identifies the customer's expectations.[12] Only with an understanding of the customer's expectations can the company or transport authority successfully define their own goals in intended service quality. Measurement of customer satisfaction thus presents the starting point for constructing quality management. Furthermore, it gains a high degree of importance through its large information content.

It is advantageous if direct fields of action and improvement proposals can be derived from the inquiry results. Inquiries about perceived and expected quality and satisfaction using a simple grade scale are suitable only to a degree because results must always be interpreted: what action is necessary for achieving improvement? For this reason measurement practices that draw heavily on *global* satisfaction are not really suitable for the implementation of quality goals. A verbalization of the individual grade levels with textual description facilitates implementation through concrete measures, but necessitates a very involved method.

It is essential for the evaluation process to declare which relevant action area is connected with which respective quality criteria. This connection is generated for the customer through the difference between perceived and expected service quality and the importance of these respective criteria. When seen as a whole, the so-called action relevance matrix (Figure 3.4) emerges, which enables the organization of all criteria according to satisfaction and importance. An example of a possible result can be found in Figure 3.4.

The criteria can be allotted into four different fields through the service provider:

1. in cases of *high satisfaction* and high importance it means: *continue in this way*, don't slacken. The results are to be kept and developed, these criteria are positive deciding factors for the customer;
2. in cases of *high importance* and *low satisfaction* the alarm bells start ringing: these aspects can cause customers to stop (re-)purchasing the service;
3. under *high satisfaction* and *low importance* criteria appear that seem *taken for granted or of secondary importance*. Additional scoring can be done here, if it is worthwhile for customer retention: unnecessary expenditures can often be discovered in this category;
4. *low satisfaction* in areas of *low importance*, finally, characterizes the criteria whose particular *faults are accepted*. Engagement with these points should be carefully considered: is there nothing more urgent?

12 EN 13816 (2002), p. 9.

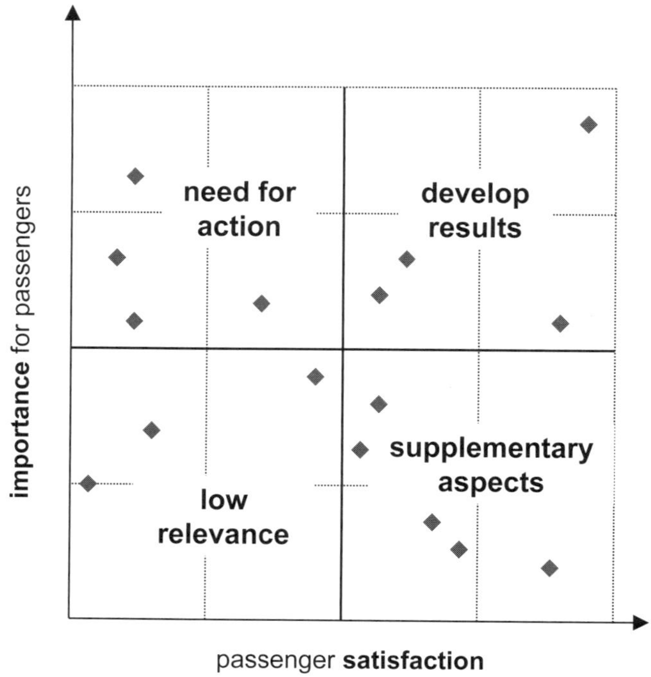

Figure 3.4 Action relevance matrix as the result of customer satisfaction

Source: author's depiction.

With the results determined in this way, the quality loop can be reinitiated: it can be tested whether objectives for intended service quality match the customer's perception and whether they should potentially be readjusted.

3.5 Users and User Organizations in Quality Management

Martin Schiefelbusch, Uwe Böhme and Hans-Peter Faas[13]

Just as much as delivering the service itself, quality management is primarily the service provider's task. Nevertheless, it makes sense to have an active dialogue with the customer in some areas (see Section 1.2.3). In the following text, possibilities for passengers to become active in this area will be presented, especially in user organizations. As already exists for participation in questions of politics and planning, involvement here can also be either formally regulated or not, initiated by customer organizations or by companies, carried out independently or in collaboration. The examples make this clear. Furthermore,

13 Section 3.5.2.2.

quality management has connections to the other levels of passenger interests (see Sections 1.2.1–1.2.4 and Figure 1.1).

The possibility of obtaining and processing information about the quality of the service proves to be decisive for the representation of interests in this area. It is not to be taken for granted that user organizations – who, in the eyes of the service provider are 'outsiders' – obtain *access to* meaningful *information* about the quality of the offering. Similar to financial and market data, companies and involved authorities, partly because of competition-related or other reasons, keep this information confidential and it is only reluctantly or partially made public. In this situation, user organizations are obliged to obtain this data through political means or Freedom of Information regulations or to collect it themselves (see Section 3.5.2).

Well-rounded knowledge of the service quality is, however, important for these organizations in order to obtain information about current problems and to orient their own work around these problems. Considering the significant importance of public (co-)financing for PT, an increased degree of transparency about the effectiveness of this use of resources can also be justified politically. In international comparison, it becomes clear that approaches to this question are truly variable.

User organizations can acquire their own information about service quality by various means. First of all, an analysis of complaints and self-conducted studies should be mentioned. For these, the organization must have an adequate *capacity* available to collect, process and analyse data. Larger investigations are, for this reason, especially difficult for local, volunteer-based initiatives. Public appeals for data collection can, on the one hand, facilitate this;[14] on the other hand, such investigations will potentially be dismissed by the companies and authorities as unrepresentative. For the analysis of passenger complaints, larger, well-known organizations are at an advantage because they obtain a larger, and therefore more legitimate, response which is less influenced by specific incidents.

The following examples refer to the quality management tasks outlined in Sections 3.1–3.3. Here, however, they will be divided into two areas in being concerned with either the establishment of *quality criteria* ('what characterizes a good service?') or the *monitoring of quality* ('how good is the quality from the passenger's perspective?').

3.5.1 Participation in the Establishment of Quality Standards

Quality standards for transport services are established in certification processes, internal guidelines, or through conventions, especially in transport

14 One example of a data collection like this is the collection of delay data carried out by the arbitration board for local transport in NRW (see Schlichtungsstelle Nahverkehr (2003)).

conventions, and are announced through advertising efforts, product standards or quality promises (such as customer charters). Generally the monitoring of standards and sanctioning in cases of failure to comply (penalties or bonus malus systems) are established at the same time.

In the description of service quality and the structuring of internal quality assurance processes, standards play an increasingly large role.[15] The certification of companies or their services is widespread in France and is also applied in Germany (see Sections 3.5.1.1 and 5.5.5.1). In communicating quality standards to the public, so-called charters are used particularly in Italy and the United Kingdom (Section 3.5.1.2). Comprehensive consultations in the preparation of transport conventions take place particularly in Great Britain (Sections 3.5.1.3 and 4.4.1). Finally, the RRT conventions of Lombardy are introduced as an example of the definition of bonus malus systems and compensatory payments (Section 3.5.1.4).

3.5.1.1 Certification in France The awarding of certificates to transport companies that offer their passengers a quality-oriented PT is especially prevalent in France. Certificates in PT were first awarded in 1993. The quality seal NF Service issued by the French Association for Standardization (AFNOR) has played an important role since then. This seal was conferred to PT in 1997; the certification practice was defined in the XP X 50 805 standard. It aims to focus the service on the customer. Information from AFNOR which is comprised of representatives from transport authorities, transport companies, passengers and responsible ministries, defined the rules and regulations for the awarding of certificates. The most important criteria are also published in official bulletins.

As a result, an intended level and a threshold of unacceptable deficient service are defined for every quality criterion. Also stated is the proportion of customers who are able to use the service quality described. Transport conventions contain clauses that state how high the NF quality should be. The choice of quality criteria is carried out on-site, which makes the consideration of some imperative (see the example in Box 3.1).

There are three types of criteria to be distinguished in this certification:

1. mandatory criteria which must be adhered to (for example information on train stations and in vehicles, punctuality, availability of lifts and escalators, friendliness of the personnel);
2. further criteria which are determined according to the local conditions (for cases in Lyon, preliminary information for disturbances, handling of grievances, presence of security and supervision personnel);
3. additional criteria as approaches to better customer service (in Lyon, the enforcement of the smoking ban, condition of transfer facilities).

15 Becker et al. (2003), p. 30; Miller and Norbert (2005).

Box 3.1

Example: Certification of transport providers in Lyon

The contract for the years 1999 until 2004 between the transport authority in the Lyon district (SYTRAL) and the transport company SLTC establishes that transport services with 'NF service' quality must be provided to 80 percent of the customers until the end of the contractual period (2004). The transport services are certified by line and as the case may be for respective delimitable operational units.

All criteria are components of an official quality handbook, which is agreed upon by a committee composed of the transport authority, the transport company and the passenger association DARLY (which belongs to the nationwide passenger federation FNAUT). The handbook will be transferred to AFNOR and forms the foundation for the following reviews of adherence to the quality criteria.

From the time that the contract comes into effect, the proportion of certified services will increase incrementally. On 1 January 2003, 77 per cent of the offering was already certified. This quality verification is seen as a reason that the passenger count was able to achieve an increase of 12 per cent in the last three years in the transport district Lyon.

Source: Compiled and summarized according to DTT (2003), pp. 77 et seq.

3.5.1.2 Customer charters in Germany and Europe Customer charters are also referred to as quality promises (see Section 2.4.2) and should publicly communicate service and/or quality standards set by the company. In addition they are also often effective internally in that they are intended to provide the employees motivation as well as guidelines for acting in a customer-oriented way.

These various motivations have lead in recent years to an increasing prevalence of customer charters – not only in the transport field.[16] They were first developed in the early 1990s in the United Kingdom where they were meant especially to encourage better customer-orientation in the area of public services (services non-transferable to competition).[17] In Italy[18] and Greece, similar considerations led to political initiatives that required, among other things, PT companies to engage in a charter. In other cases they are initiated by the company itself as a marketing measure. In recent years, European umbrella

16 Prevalence and design of value pledges in European transport see *in extenso* the EUSG consortium (2006).

17 See Pinney (1993); Prosperetti et al. (2000), pp. 17 et seq.

18 In the United Kingdom a charter is only requisite in the regulated part of the transport market (see Section 5.7). The following statements refer only to this area.

organizations of transport companies have played an important role as well in the introduction of such documents and in the creation of templates that are implemented in degrees by their members.[19] The political passengers' rights discussion should not be overlooked (see Section 2.6) in which (voluntary) commitment to a charter is propagated as an alternative to legal provision. This is true as well for the customer charters of the DB AG which present the best-known case of this kind of document in Germany.

Possibilities for participation in the formation of a charter are likewise formally regulated in the United Kingdom and Italy in that user organizations get the possibility to give comments on text drafts. The content of the document as well is in part explicitly regulated (especially in Italy, see Section 5.6.4) while otherwise there usually exists a large leeway in the shaping of the content's scope as well as its form in the pledge. According to the parameters of PT service, there are however clear focuses, at least in the topics addressed, in the area of punctuality and customer service (see Figure 3.5)

While in service guarantees,[20] the offered compensation and requirements for claims have a decisive and economic significance, this is much less the case in a charter, which indeed includes no kind of obligation for compensation. For this reason, a document of this kind should at least provide a comprehensive and comprehensible overview of the quality offering. Figure 3.6 shows the customer charter of the Schwäbisch-Hall city bus, which in this sense can be taken as an example of a comprehensive charter.[21] Further examples for the formulation of customer charters are contained in the sections about the United Kingdom (Section 5.7) and Italy (Section 5.6).

A more detailed examination of the effects of charters has only taken place to a limited degree to date in part because they have only just been introduced. The intended internal and external effects are, furthermore, often not directly verifiable. The perception so far from the view of the user, in contrast, highlights a few critical points. The content and style of the document have thus been criticized as imprecise and non-binding. For concrete passenger problems, the announced standards are generally not helpful. The publication of the charters is also seen as insufficient.[22]

3.5.1.3 Transport service contracts: participation in Great Britain and Germany In Great Britain as well as in Germany, transport services in train transport are allocated through a call for tenders and regulated by service

19 These are mainly the charters for the railway transport of the European railroad associations (passed 2002, see e.g. CER (2005)) and the charters of the UITP (2006). Similar documents exist in long-distance buses and air transport (see EUSG Consortium 2006)).

20 See Section 2.4.4 and Neugebauer (2007).

21 See EUSG Consortium (2006), pp. 83–6.

22 For more detail on this see EUSG Consortium (2006), esp. pp. 121–2.

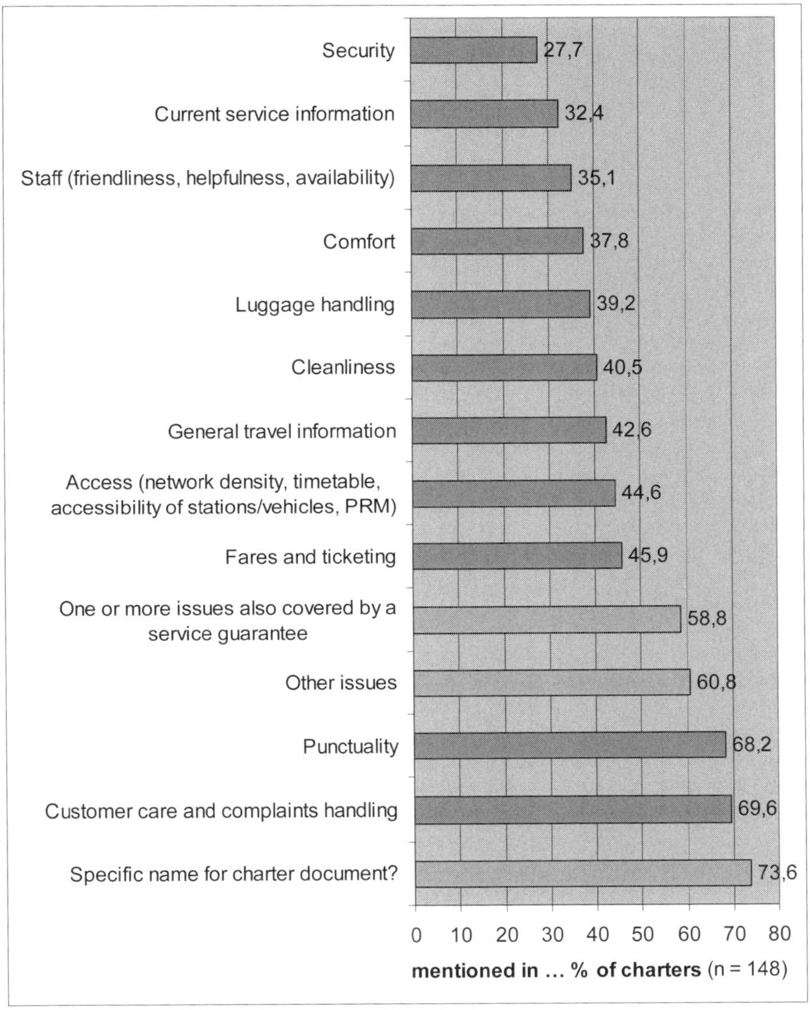

Figure 3.5 Contents of European customer charters: incidence of relevant parameters

Source: EUSG Konsortium (2006)

contracts (for more detail on this see Section 4.4.3). The user's possibilities for participation, however, prove to be highly varied. While comprehensive consultations with the associations take place in Great Britain, in Germany the user organizations are only able to contribute informal input. Both contrasting forms of participation are presented in the following sections.

Dear Passenger
Our goal is to make every trip with us a pleasant experience. And because we are not content with spoken words, we have defined the criteria for the quality of our performance in a passenger charter. In these ten commitments, you will recognize that your well-being is close to our heart. We will tell you how we will ensure that everything will run even more smoothly – and what we offer you should it happen that you are not entirely satisfied despite our attention.
We are happy to be there for you!

Local Marketability
To live in Schwäbisch Hall means to live with the city bus. Have you ever been shopping with us? Unload your shopping bags – or whatever may be burdening you – in our Customer Centre at Spitalbach in the heart of the city.

Information
Nothing is worse than not knowing what is going on. With us you will have not only your timetable in pocket format and easily readable and understandable notices, but we promise as well to keep you constantly updated with extensive current information.

Punctuality
We do not need to guarantee it – we are it: the Schwäbisch Hall city bus is especially punctual according to 2002 passenger surveys. If, however, it should happen that we pick you up too late, we will pay you back with a cinema gift coupon.*

Complaint Management
Come talk to us! We will see to it that your concerns are handled promptly and in a friendly manner. We are thankful for hints, because they allow us to improve. We promise that we will get an answer to you within three workdays.

Comfort
Is your seat comfortable? Is the air on the bus top-notch? Can you answer both with a yes? We use primarily low-floor buses with climate control.

Cleanliness
Crumbs in the car, shoes on the sofa, drink containers under the bed... Do you do that at home? The city bus endeavours daily to keep the vehicles neat and clean for you. Please let us know if we have overlooked anything.

Safety
We start early! 'Bus tips for kids' is the name of the brochure for our little ones (but not only them), to learn how to safely interact with our buses. It goes without saying that our drivers are fit and, if it should be necessary, will actively step in for your safety.

Clarity
Together with the transport association, we make fares and timetables easily understandable and clear for you. Is something still unclear for you? Then just ask your driver. He will give you information about the fastest connections, your transfer possibilities and the cheapest fares.

Standard and Continued Training
Are you the exam type? Not taking part just won't do! Our drivers and associates have to undergo our exams at regular intervals and take part in training courses.

Environmental Friendliness
The newest motor technologies, filtration systems, fuel-saving operation methods and process water for vehicle cleaning: in procurement and in everyday use, we try to satisfy the set environmental standards. Help contribute to the bus's continued gain in importance as an environmentally friendly mode of transportation in Schwäbisch Hall.
Climb onboard!

Figure 3.6 Customer charter for the city bus of Schwäbisch Hall

Source: Stadtbus Schwäbisch Hall.

Great Britain: As soon as details about the content of the contract between the contract winner and the contracting body are known, there exists, in principle, the possibility to give an opinion about it. This is the site of the actual focus of a passenger organization's activities. The dissemination of bad news about the service of a transport company ultimately has only negative consequences for the later chances of winning a contract on the British (rail) market. Because many companies are internationally active, the consequences would be felt in other transport markets as well. In addition to this leverage, Passenger Focus (PF) in particular, which acts as a watchdog group (see Sections 4.6.5, 5.7.4.4) can according to law report to the Department for Transport (DfT) about the transport company's contractual violations, which can result in punitive payments. PF itself cannot, however, make decisions in this regard, but acts only as in a consultative capacity.

Every train transport company's operation is subject to constant checks to test whether they are adhering to the contents of the contract. This checking consists of, for example, performance figures for punctuality and reliability, and of complaint management (response time for complaints) for which non-adherence incurs malus payments. It is, however, decisive for the passenger that companies are required to report these payments to the DfT as well as PF. In this way nationwide statistics[23] become public at regular intervals, statistics which also show a comparison to the previous charges. A lasting decline of general service thus becomes visible to everyone, which offers companies an incentive to provide good performance.

Germany: After signing transport contracts, months or years usually pass before operation actually begins, as vehicles must be built or employees trained according to the contractual requirements. This time can be used by user organizations to remedy difficulties that have come to be known. The main activity, however, is doubtless the lobbying work that takes place *while* the respective transport company is already offering their transport service. It is only in this phase that the actual deficits and potentials for readjustment start to be shown, to which organizations and associations can draw attention through publicity in relevant media such as their own organization's periodicals, trade journals and newspapers.

However, the experiences of German passenger organizations (see Section 4.6) show that changes does not have to be achieved through media pressure at first. Often, cooperation takes place on informal levels[24] where ideas and differences in opinion can be exchanged. This way can in part lead to faster, less complicated and more effective results than formal participation possibilities. An example of this would be the work of passenger advisory boards, which reside directly with the respective policy-making bodies. This

23 See Office of Rail Regulation (2006) as well as Passenger Focus (2006)

24 Contacts through early conjoint work, meetings at conferences etc.

proximity can foster consideration of proposals; the advisory boards only meet, however, at most a few times per year.[25]

3.5.1.4 Incentives and sanctions: rail transport contract Lombardy region, Italy[26]

Concept of the contract: The transport contracts which have been held between the region of Lombardy and the transport companies for RRT services contain comprehensive descriptions of the service quality to be provided. Flanking this, a bonus malus system has been developed that builds on customer perspectives and is meant as well to give the travellers a participatory role in the mali to be paid by the company.

The contract for the years 2004–2005 comprises a total of 13 quality criteria which are presented in Table 3.1. The monitoring of delivered quality is carried out primarily through test rides and on-site observations through a contracted institution (see Section 3.4); information about train cancellations, delays and capacity reductions are communicated monthly by the transport and infrastructure companies.

Non-adherence to quality standards is linked to malus payments (fines). In addition to the maximum amounts per quality criterion, a limit for total payment is stipulated which, in the case of Trenitalia, amounts to 4 per cent of the value of the contract in 2007 (2004 still 2 per cent).

Customer interests in the transport contract: Customer satisfaction is not included as a criterion in the bonus malus account. Nonetheless, the contract addresses the customers' views at a few points. The company is required to process *complaints* within stated deadlines and to report to the region monthly about the type and extent of the complaints. Non-compliance with this requirement results in a malus payment. The region has the right to make inspections of the correspondence, and in cases where it forwards complaints to the TC for processing, receives the answer in duplicate.

Moreover, the company, together with the transport authority, is required to organize *regular meetings* with user organizations and commuter groups, to contribute constructively to these and to make yearly updates to the Carta dei Servizi (see Section 5.6.4) – in accordance with national parameters – according to consultations with user organizations. In addition, the region has recently been including the organizations in consultations about how the malus payments should be put to use.

---

25 See Section 4.6.3 and Meyer-Liesenfeld (1997), p. 5

26 This section is based on materials found at http://www.trasporti.regione.lombardia.it (informazioni tecniche > organizzazioni e competenze > servizio ferroviario regionale). For more extensive information, I should like to thank the Ufficio Servizio Ferroviario Regionale – Regione Lombardia.

Table 3.1 Quality criteria and malus payments[a] in the Trenitalia transport contract

Quality criteria	Indicators	Max. possible fine (€)
Reliability	Index (see Fehler! Verweisquelle konnte nicht gefunden werden.)	Payment as bonus (see below)[c]
Punctuality	Monthly average of delays > 10 min	1,000 per train run
Repeated cancellation in train runs[d]	More than three cancellations in one train run per month (also for bus substitutes)	1,000 per train run
Route closure	Cancellations > 50 % of trains run on one route in 15 days	100,000 per incident
Total cancellations	Cancellations > 0.5 % of transport service p.a. Cancellations > 0.6 % of transport service p.a.	200,000 additional 300,000
Insufficient seating	Falling short of stipulated seating offering in the peak hours in at least 90% of monthly transport days	1 per seat
Overcrowding	Failure to accommodate the demand with capacity (overcrowding of offered seating determined in two surveys per year)	1 per seat
Condition of tolling stock[b]	Functionality of doors, toilets, heating, Lighting (if applicable also climate control and loudspeakers)	Max. 400,000 p.a. class B[e] Max. 800,000 p.a. class C
Cleanliness of trains[b]	Appraisal of interior and exterior, and non-Adherence to cleaning dates *Additionally*: neglect of base cleaning (acc. to reports from the TC)	Max. 400,000 p.a. class B[e] Max. 800,000 p.a. class C 50 to 1,100 per defaulted cleaning (according to extent)
Cleanliness of train stations[b]	Appraisal of interior and exterior train facilities	Max. 200,000 p.a. class B[e] Max. 300,000 p.a. class C
Information in train stations[b, f]	Equipment in train stations with determined information elements *Additionally:* delayed repair of electronic information media	Max. 100,000 p.a. class B[e] Max. 300,000 p.a. class C 100 per day (after 3 work days) plus 1,000 if determined by transport authority
Distribution system	Lack of stipulated distribution channels Functionality of distribution channels	2,000 per incident 100 per lost day plus 1,000 if determined by transport authority
Complaint management	Non-compliance with monthly requirement to report to transport authorities	500 per incident

Notes: a = The contract contains bonus payments as well for certain criteria which, due to space, are not detailed here; b = survey through test rides with categorization in three condition classes (A = no complaints, B = some complaints, C = critical condition); c = final amount dependant on proportion of deficient rides; d = fines for cancellations are understood as additive and additional to fees retained by purchaser. Only cancellations for which the TC is responsible are to be paid (not including force majeure, planned building work and strikes); e = for routes with high density of trains no bonus for passengers, instead a fine of €5,000 to the region per exceeding of monthly index value; f = for quality deficiency in the area of train stations, fines will be charged but not levied because as yet no contract between transport authorities and infrastructure companies exists.

Source: Regione Lombardia 2004.

A stipulation for a *compensation* (bonus) for noncompliance with punctuality standards has been in place since 2003 addressing the direct interest of the passengers. For this, a monthly, route-based index (Figure 3.7) has been established which repays owners of month and year passes with a discount of 10 or 20 per cent off renewal of their ticket. The index takes into account delayed and partially or completely missing trips as well, but not missed connections.

$$\text{Index} = \frac{\sum \text{delay minutes} + \sum \text{non-arrival minutes (without uncontrollable events)}}{\sum \text{timetable minutes} - \sum \text{delay minutes (uncontrollable events)}} \ [\%]$$

Definitions:

Minutes of delay:	Delay at the end station of more than 5 min. without including "uncontrollable events" (according to stipulations in criteria catalogue)
Minutes in the timetable:	Planned train travel times
Non-arrival minutes:	Train trips that did not take place and were not replaced with another form of transport within 60 min. of the missing trips

Figure 3.7 Account of the index of reliability in the RRT of Lombardy

Source: author's depiction.

The compensations to be paid are financed by the malus payments – in 2004, approximately €595,000 was credited to commuters, while the region withheld an additional c. €2.9 million in fines. In 2005, these values rose to €685,000 and €3.6 million respectively.

The bonus was the first such compensation in Italian RRT, in the meantime other such offers have been introduced in other regions. The regulations in their details do not always seem equitable to the passengers, as the letters to customer magazines show. The bonus is seen as (partial) compensation, but due to considerable quality deficiencies have not so far been able to provide a lasting contribution to satisfaction. Nevertheless, the concept shows that payment of compensation to passengers is definitely compatible with quality-contingent compensation from transport authorities.[27]

27 See also BAG-SPNV [FAPA-RRT] (2006).

3.5.2 Participation in Quality Checks

User organizations, as earlier presented, can access or collect their own data for the monitoring of service quality with the partial participation of a third party or on their own. This can happen by means of satisfaction surveys (see Section 3.4), other polls or inquiries, or test rides. This brings with it the opportunity for passengers to be involved in quality checks on individual levels as quality scouts, train station guardians or quality testers. A similar function is ultimately fulfilled by the complaint management team, in which passenger organizations occasionally participate (Section 2.3).

Some collections of data about service quality are carried out in highly various ways by passenger and user organizations. To date in Germany, the environmental association VCD[28] and the arbitration board for local transport in the country of NRW have carried out major inquiries. Another important stakeholder in this field is the Stiftung Warentest. In the other countries considered here, a few Italian user organizations were acquainted with studies that were likewise mostly co-financed through third parties. In contrast, the Chamber of Labour in Austria (see Section 5.1.4.1) and the British Passenger Agency Board have comparatively much more substantial resources of their own – their participation possibilities in quality checks will be discussed in detail below.

As an example for individual participation the Berliner Verkehrsbetriebe (BVG)[29] will be presented in Section 3.5.2.2, which was introduced in 1996 as perhaps the first of its kind. Inferences about quality can also be made alongside the typical practices of satisfaction measurement (Section 3.4) based on the handling of individual cases (Section 3.5.2.3). Accounts of other concepts are collected in the literature.[30]

3.5.2.1 Quality reporting in British public transport In the United Kingdom numerous data about quality are published for regulated PT (railways, Greater London, Northern Ireland, see Section 5.7) through transport authorities and agencies. Accordingly, passenger advisory boards such as London TravelWatch or Passenger Focus have always had the right to access these documents; they are also, of course, posted on the Internet. The secretariat of the passenger advisory board uses this information to oversee service quality and to produce comprehensive reports for its members. The public has access to these as well,

28 Such as the Train Customer Barometer [*Bahnkundenbarometer*], which has been performed several times, and the contest 'Regal Affairs in Bus and Train'. [Königliche Verhältnisse in Bus und Bahn].

29 See also Buschmann and Grunwald (1998).

30 Schrameyer and Frittgen (1999); Schürmann (2003); Krohn (2007).

and they are, along with the information collected by the organization itself (see below) also used in the organization's public relations.

Quality monitoring and analysis done by these committees comprises 'hard' as well as 'soft' factors of customer satisfaction, for example, cleanliness, information and the personnel's helpfulness. For aspects that are not covered in publicized statistics, they can carry out their own inquiries and studies, or hire third parties to do so. Further material is found in passenger complaints which the committee examines (see Section 2.5). Among studies carried out by the organization itself, the National Passenger Survey (NPS) should be pointed out. This survey deals with regular measurement of satisfaction among train travellers and has been carried out all over the country since 1999 four times per year with approximately 25,000 inquiries each.[31] In addition to a regular analysis of the overall situation, the data allow comparisons between the individual rail communities because the analyses generally occur on the company level. After the reorganization of the passenger advisory board on the one hand, and the railway administrations on the other, Passenger Focus has been responsible for carrying out the NPS since the middle of 2005.

3.5.2.2 BVG Quality Testers The Berlin Quality Testers are LPT customers who volunteer free of charge to answer questions about the performance and service offerings of the Berlin Public Transport (BVG).

Since 1996, Quality Testers have generally been asked to comment on determined problems several times a year. Particularly for written inquiries this group, which consists of c. 170 people (as of 2007) presents a quickly initiated customer sample. Quality Testers are, however, also very well-suited for questions with more depth, which can only be determined within a group discussion. For these reasons, Quality Testers are customers who can be easily and quickly addressed for qualitative and quantitative studies, that is group discussions, standardized interviews as well as inquiries as preliminary studies for planned research.

These processes deal with questions and evaluations:

- about performance and quality features of individual routes or groups of routes;
- about facts that otherwise would or could not be inquired into in reference to extant facilities, vehicles, operating procedures and service offerings (innovations, differentiations, diversifications);
- about additional topics such as changes in fare structures or new generations of vehicles.

Advantages of the Quality Tester concept are:

31 Methodic commentary at http://www.passengerfocus.org.uk/your-experiences/content.asp?dsid=426 (accessed May 2007).

- prompt, regular and cost-effective measurement of customer needs;
- especially with 'soft' quality features such as cleanliness and safety, this process can close gaps in knowledge. With such features, there is a multiplicity of perceptual dimensions, which can only be partially completed by standardized practices due to parameters that cannot be determined;
- data collection possibilities that can be used at short notice to determine customer attitudes about current questions, such as changes in the LPT offering, which enables a better understanding of customers' future demands;
- identification of concrete fields of action for quality improvement measures and determination of priorities;
- a signal from the BVG that it is interested in the opinions and attitudes of its customers;
- the BVG's Quality Testers act as positive disseminators with their activities and experiences ('The BVG is doing something for its customers, and we're helping them out').

The Quality Tester's customer group is, however, not representative. Almost all of the Quality Testers are regular customers who use the BVG at least once a week. Men are over-represented (70 per cent), people over 60 are also more represented than the average regular customer. Thus, the Quality Testers do not present a representative picture of LPT customers. The information gained from the Quality Testers can therefore not fulfil the requirements for representative data.

Quality Testers work as volunteers, that is unpaid. In regard to their needs as regular users, 40 per cent of Quality Testers list 'transport topics' as a hobby. Quality Testers are courted with material and conceptual incentives such as small gifts or invitations to events, parties and depot visits. It has been determined that it is especially important for Quality Testers that the members of the board are available from time to time for personal contacts at events. The desire expressed by some Quality Testers for free rides in BVG transport has not been fulfilled due to the danger of it resulting in such things as a too-positive judgement of the BVG.

In order to ensure the Testers' function as disseminators, it is necessary to send them regular background information. It has been determined that it is helpful to give explanations of the histories about technologies, operations, transport, transport politics and general societal themes. This helps to increase understanding in terms of individual things seen as LPT quality deficiencies. In principle, results should be communicated to Quality Testers in a suitable manner after the conclusion of each action. This can help create a positive dissemination effect.

The BVG's Quality Tester concept can be an essential element of customer orientation with according maintenance and intra-corporate acceptance. In this respect, the costs justify themselves including the preparation and sending of surveys, data entry and analysis, reporting systems, care and incentives, because the understanding that is gained can be effectively used for improving quality in performance and service offerings.

Above and beyond the exchange of standardized surveys, this concept creates an active dialogue with the Quality Testers. Suitable communication and implementation capacities must be taken into account by the company.

Recent years have shown a strong excess of elderly participants and a strongly increasing proportion of men in this group. This situation leads to problems in terms of representative conclusions and quotations for group discussions. While non-representative studies can for certain reasons be a goal of market and opinion research, it makes sense to take all aspects of customer groups into account. For this reason, it would behove the BVG to spend more energy addressing young and female customers in order to be able to continue to represent the concerns of these customer groups.

3.5.2.3 Quality checks on the single-case level At least a short consideration should be given to opportunities for passenger and user organization activities for dealing with concrete passenger problems and complaints. As the introduction of this chapter already mentioned, a content-based examination on the single-case level necessitates working capacities that quickly exceed the possibilities of many smaller organizations. On the other hand, many citizens at the same time expect user organizations to provide advice and help on a single-case level, such that it may rarely be possible to completely avoid these questions. Simple instruments are therefore essential. Examples of this are passenger consultation days and suggestion cards.

Passenger consultation days are public events at which representatives of the transport company and of the organization work together to address questions, concerns and criticisms from passengers. They use the occasion to answer/respond when they can, but a later reply can also be arranged. A public discussion as a core practice can be supplemented with an introductory speech and/or bilateral interview possibilities. This allows such events to offer an occasion for direct contact from which participants can also profit afterwards. They also have news value and can strengthen the company's as well as the organization's profile. Consultation days following this concept have been carried out by the Berlin passenger organization IGEB as well as, in a modified form, the British organization, Bus Users UK.[32]

Suggestion cards are postcards with pre-printed areas on which travel experiences and problems can be briefly described and sent to the organization (Figure 3.8). Due to their pre-structuring, filling out the cards is easier than

32 E.g. passenger consultation day for regional rail transport: http://www.igeb.org/pressedienst/ files/igebpresse_20060926.pdf; Bus Users' Surgery from Bus Users UK: http://www.bususers.org/ about.htm (both accessed July 2007). See also Section 5.7.4.3).

writing a freely formulated letter and their processing is simplified. Because a certain cost expenditure is however required, the organization should make clear from the beginning whether the offer is to be understood as individual consultation or just as an outlet. The fundamental idea behind suggestion cards corresponds to that of a complaint form as many transport companies already make in paper/card or electronic form. Its origins date back to an idea from the Berlin passenger organization IGEB in the 1980s.

The organizations that have access to lasting funding and according human resources pools also have significantly favourable conditions for processing incoming suggestions. This applies again to the British user advisory boards and to the French and Italian organizations as well whose activities also comprise mediating in cases of dispute.[33]

3.6 Quality Management between Experience and Science

Martin Schiefelbusch[34]

Public transport is, like all services, not tangible; a deficient service can thus not be repaired, but is only replaced by another (such as a later departure instead of one at the planned time). Whether the service can be delivered as planned depends not only on the company, but on external influences as well. The generally accepted definition of quality (see Section 3.2) does not differentiate whether external or internal causes inhibit the adherence to stated standards. It thus takes on the customer's perspective for whom this distinction is insignificant (see Sections 1.2.4, 2.4.4). For this reason, quality management must consider both aspects.

At the centre of these considerations is the service *provider* – it is they who must meet demands in order to provide their service with the proposed quality. The exact description of input, processing and intended results plays a central role in this as well as documentation and regular target/performance comparisons.

On the one hand, quality management ultimately serves *customer* satisfaction, on the other hand, the technical/organizational aspects that accompany this often seem abstract and less suitable for lay participation. It is thus no wonder that the forms of customer participation (see Section 3.5.1) described here are predominantly oriented toward user *organizations* – organizations that have at least enough capacity and resources to become acquainted with questions of complex detail. An effective co-determination seems to demand this as well. In its own way, dialogue with laymen also

33 See Sec. 2.5 and Schiefelbusch (2007)

34 I would like to thank Hartwig Meier and Nancy Neugebauer for their helpful suggestions.

Figure 3.8 'Grievance card' from the Berlin passengers' organization IGEB

Source: IGEB.

warrants a realization of their expectations. Participation of this kind is however, quite obviously a question of strategy and market research.

The standards described in this chapter regulate quality management in PT only on a basic level and thus leave free space for a design that is fitted to site-specific conditions. All the more careful attention should thus be paid to choice of indicators and practices on-site. It is important to encourage neither an aggregation that is too strong – limited too much by global satisfaction to infer concrete action areas – nor to collect mountains of data which ultimately cannot be evaluated with available resources. Institutional arrangements should be considered in this as well: on the one hand, competition on the market without doubt contributes to giving quality management significant importance. On the other hand, this very competition tends to make companies notably reluctant to deal with (not only) quality data. This speaks for giving a larger role to transport authorities and possibly user organizations as well as cross-company institutions.

In service provision, the relationship between abstract quality terms and the users' concrete experience becomes more evident. This is not only a matter of learning from customer criticism as discussed in Chapter 2, but also of maintaining the standards reached beyond individual problems. If quality is defined as fulfilling previously defined criteria, it is not only necessary to determine what these criteria are, but also to document to what degree they will, in real terms, be reached. This subsequently is of interest to the company which must draw its conclusions based on this determination, and to the transport authorities and customers who both finance the service and would like to know what they have to prepare for when using the service.

Information about delivered quality is to date, however, not accessible everywhere, especially because companies fear the negative effect of poor data, and the fact that numbers without background information can be easily wrongly interpreted. Here as well the quality loop shows another potential perspective: lacking data about delivered quality, perceived quality and customer satisfaction are accordingly more strongly based on personal experience, on the one hand, and opinions and prejudices, on the other. The latter in particular (see Section 2.2.2) are often shaped by negative experiences, even if these only refer to a small part of delivered service. An increased degree of transparency about actual affairs could contribute to putting such impressions into perspective – it would in any case be worth a try.

Chapter 4
User Participation in Public Transport

4.1 Introduction: Starting Points for Participation
Christian Bodensteiner

The participation of customers in the planning of public transport is the topic of this chapter. The term 'participation' and the synonymously used term 'involvement' are generally defined as the incorporation of citizens in decision-making processes.

Participation can occur on four levels: politics, planning, production, and practice (see Section 1.2). On the political level, numerous kinds of involvement are possible. Section 4.2 explores in further detail possibilities for citizens to influence the politics of public transport issues.

The participation of citizens and/or customers is also possible on the planning level. Section 4.3 first gives an overview of the history, terms, characteristics, and goals of participation. Section 4.4 goes on to examine the extent to which local public transport, with specific regard to the organization of public transport networks and timetables, makes legally defined 'formal' possibilities available to the public, and the planning sessions in which these are used.

Section 4.5 introduces examples of participatory practices that have been tested and conducted in an institutionalized setting. In general, these practices are highly cooperative and informal (see Sections 4.3.2 and 4.5). Other interest representation groups include boards and institutions such as passenger advisories and organizations that are discussed in Section 4.6, before Section 4.7 briefly covers the conditions for success and the limits of participation.

4.2 Participation on the Political Level
Martin Schiefelbusch[1]

There are close links between transport politics and transport planning, and it is often difficult to clearly distinguish them (see Section 1.2). Therefore, many of the planning and design issues discussed in this chapter *also* have a political dimension, even if it is not explicitly mentioned. On the other hand, the forms

---

1 For their helpful suggestions in this section, I should like to thank Christian Bodensteiner, Holger Jansen and Heike Walk.

of 'regulated' participation described in the following only cover part of the (transport) political discussion and decision-making process.

The following section is thus meant to address the formation of 'political' opinion in a narrow sense. In the context of this book, there is not room to define 'how politics work'. Rather, there will be a brief discussion of most important instruments that can be employed in shaping the process. The description is oriented toward the perspectives and possibilities of volunteer organizations in the field of transport, but can perhaps provide helpful suggestions for other groups and issues.

4.2.1 Politics as a Process

Policy-making takes place in a cycle (Figure 4.1), which among other things is fuelled by experiencing the contemporary situation (perception and definition of issues). In the response to this situation (from the development of solutions to political study), various political powers almost always have differences in how they perceive and assess the situation and the possibilities for action, because they are conditioned by essentially differing concepts. From this situation emerge three fundamental starting points for influencing the policy-making process: the perception of issues, the development and assessment of solutions, and the exercise of influence over fundamental values.

In light of the differing perspectives, policy-making is ultimately a negotiation process in which consensus and/or majorities must be found. In this process, both power relations and communicative abilities play a role for those involved. This process takes place in certain *formal steps* – such as parliamentary debates and hearings – but also outside of such practices in *informal* dialogue with and between those involved. For successful cooperation it is necessary to be aware of *both* areas and, as much as possible, to make use of both.

Organizations that represent passengers' interests usually do this work across all parties. Their members' affiliations with parties or parliaments are likewise contingent and play a subordinate role in the organization's work as a representative of group interests. Passenger organizations are thus outside of narrow political systems (delegates, administrative officials and members of parliament).

This creates a fundamental challenge in that, on the one hand, these organizations want to play an active role in this system, and on the other they are included in its processes only to a limited degree. Informal dialogue – lobbying – thus takes on a special significance. In the following section, a few political practices that allow 'outsiders' to play an active role will first be introduced before the section following that addresses informal participation possibilities (lobbying work).

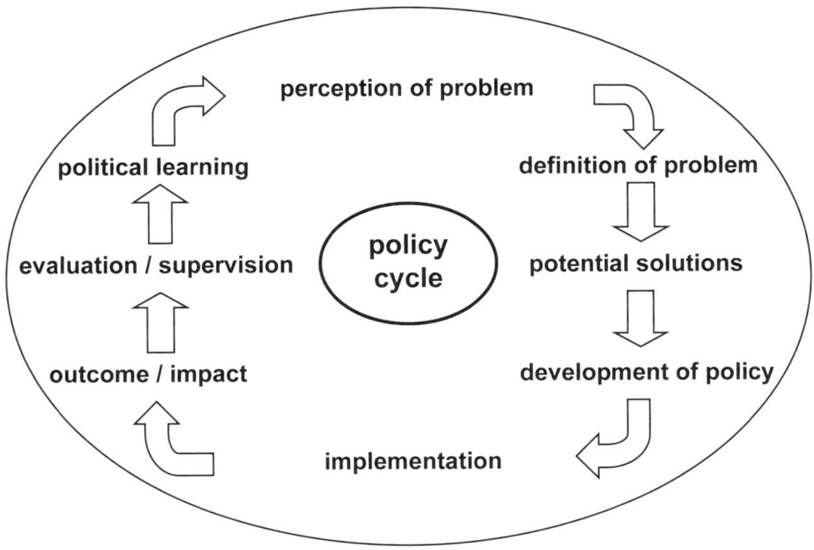

Figure 4.1 The policy cycle

Source: Tanzler et al. (2005), p. 5, based on Howlett and Ramesh (1995).

4.2.2 Participation in Political Practices

There is a central significance accorded to the parliament and the administration in the democratic political process. The ideal model of an official decision-making process free of external influences is, however, not significantly altered by lobbying alone; parliamentary practices themselves also offer possibilities for 'controlled' exercise of influence by interest groups.

The following list is intended neither to provide a complete picture, nor to create the impression that the listed possibilities are available everywhere without limitation. Rather, each one must be considered in light of the differences in regulation among the various levels of government (federal, state, municipal), and the respective ordinances that apply in various regions (such as municipal codes in federal states and regulation of citizen initiatives).

- *Petitions* are a basic right which individuals and organizations alike can make use of. This assures that concerns will be discussed by the responsible parliamentary bodies; this includes research and consultations with other stakeholders. Petitions are often seen as a 'last resort', when other forms of exercising influence are not successful. Public advocates serve a similar function, such as advocates for the concerns of specific groups.[2]

2 Cf. Hopp (1992) and Section 2.5.

- *Petitions for referenda* and public decisions can in principle also be used for transport politics. Along with the limitations set forth in legal guidelines, it should also be considered that only parliament, government, and administration are the potential recipients of such requests. Transport companies or administrative unions, which are legal entities of their own, can only be addressed indirectly as the executives are required to respond with specific actions.

  Collecting the necessary signatures often requires a substantial commitment, which can only be achieved by accordingly equipped citizens' initiatives or organizations. This instrument is best suited for questions of fundamental principle or other issues that expect widespread interest from the public.
- Recognition as an *expert citizen* offers an individual (who can also belong to an interest group) the opportunity to have their voice heard in the political process as an advisor. As opposed to informal exchange, this position includes a right to speak and/or be invited to enter into debates to represent specialist knowledge or positions. This is especially interesting on the regional level because, on the one hand, separate affiliation with specific issues is less common, and on the other, policy making often depends on volunteers, who have less access to specialist knowledge and research than higher levels.
- Participation in parliamentary *hearings*, in contrast, offers organizations the opportunity to represent their positions. Such an invitation should certainly be taken advantage of, particularly because such instances also have a good effect on public relations and offer the opportunity to have lasting contacts. Hearings take place in the context of planning practices as well (for more information see Sections 4.3 and 4.4). In order to be invited, however, organizations must be officially recognized for previous advocacy work (see above).
- *Parliamentary requests* can generally only be posed by Members of Parliament or political parties. If such contacts exist, however, external organizations can certainly provide impetus for such requests. Parliamentarians can use this opportunity to distinguish themselves, if the issue touches upon a current or particularly relevant issue. It is not uncommon that requests have a controlling function for administrations, for whom grievances (such as sluggish implementation of political decisions) are brought to bear. This practice also offers an interesting opportunity for the gathering and publicizing of information that would otherwise not happen, or only at great expense.
- Participation in *advisory boards* offers a lasting and therefore relatively comprehensive opportunity to have a say in political issues. The most common type in public transport are so-called 'passenger advisory boards' (see Section 4.6.3), which are often established at transport companies, though they can sometimes be found with transport

authorities and municipalities. Other advisory boards exist in a wide variety of organizations, in which passenger and/or consumer representatives, among others, can take part.

Independent of the direct participation of passenger organizations, these practices can be useful as sources of information through documentation (in press releases, annual reports, the minutes produced at meetings, etc.), which is important for informal lobbying work.

4.2.3 Characteristics of Lobbying[3]

The general public often perceives lobbying as negative and associates it with the exertion of pressure toward implementing 'self-serving' ends and even with corruption. However, expressing concerns to policy makers is, first of all, a legitimate practice and cannot be avoided in a democracy, even outside of the party system. For this reason, it makes more sense to have a broad awareness about lobbying activities and to engage in open discussion about its necessary limits than to make it into a taboo.

The main tool of lobbying is *informal dialogue*, which does not have to be limited to delegates, but should include both their associates as well as the associates of people involved in administration (ministers, regional councils, administrative unions, and so on). Specialized knowledge is very helpful in this endeavour in finding the right contacts through the network of dispersed responsibility. In view of the significantly better financial means that companies tend to have, it is important for organizations to have an even more efficient allocation of resources.

The second main feature is *public relations*, for which a number of tools are available: press releases, press conferences, studies and other detailed documents, collections of signatures, demonstrations and other similar campaigns. The common goal of these activities is ultimately to garner public attention, to win support and thus influence the political decision-making process. The two areas are hard to separate from one another. What is important is that the various instruments are used in a balanced manner.

There are unfortunately no simple rules for successful lobbying. It is ultimately a question of persuasion whose prospects for success depend on the subject matter (the quality of the arguments), the way they are communicated, and the resources available. The degree of professionalism on the side of both politicians and lobbyists varies significantly according to the different levels in question (Europe, country, state, municipality). The following aspects, however, can be sited as basic requirements:

---

3 For this section see Ebbers (2005), Kleinfeld et al. (2007), Leborgne (2005), Schuster (1996) and Anon. (2007) for examples.

- *Up-to-dateness*: issues have to be 'raised' at the right time, which can be determined by the issues themselves as well as by external conditions. This means reacting quickly to current events (such as political initiatives addressing other urgent quality deficits) and exploiting them for the resulting public attention. In order to do this, the necessary resources (spokesperson, press contact, press list) must be prepared. Other time-related parameters include the scheduling of changes in local transport timetables, planning meetings, dates for tendering bids and, of course, upcoming votes and their often lengthy fore-runnings.
- *Expertise*: issues must be substantiated with expert knowledge. It helps to thoroughly prepare one's position, to advance it and deal with potential counter-arguments. The more specific the issue, the more likely it is that the stakeholders must convey knowledge in addition to being persuasive. Having one's own expertise also helps to judge and react to the arguments brought into play by other stakeholders (transport companies, administrations).
- *Coordination*: the implementation of various instruments and the activities of individuals must be coordinated. This provides for both interior working efficiency and exterior persuasiveness. Making contradictory statements to the public and passing on confidential information should be avoided.
- *Strategic awareness*: even if the endeavour is ultimately one of persuading others of one's position and/or the importance of one's concerns, this must be done with regard to the political context. This means not only 'competitiveness' – being familiar with other opinions, alternative planning variants and their advocates, and putting oneself in their position. It is also a matter of the degree to which one wants to represent fundamental positions or make pragmatic arguments, of what should be discussed in public and what internally, and of what allies can potentially be won (see Section 4.6).
- *Character*: these requirements must ultimately be fulfilled by the people who have active roles – especially those who come into contact with policy makers and the public. Alongside tact and know-how, the ability to communicate is especially important.

4.2.4 Challenges for 'Passenger Politics'

In the everyday political work, what is described as 'eloquence' traditionally varies among lobbying organizations. While the interests of companies and employees are generally represented by well-endowed organizations (umbrella groups and/or unions), this is significantly more rare for consumer interests.[4]

---

4 See Burger (2006); Falkemark (1999), p. 312; von Hippel (2001); Müller (2001).

For public transport this is often particularly true due to the fact that passenger initiatives tend to operate on a local level. Local organizations are also at a disadvantage because many issues are decided on higher political levels, especially those that affect the financial and legal basic conditions of PT. The engagement and dedication of the membership thus play an accordingly larger role in such organizations. Another consequence is that lobbying only makes sense if lack of material endowment can be compensated for by expertise and good communication. For this reason, organizations should tap the assets and expertise that are provided by the experiences of their own membership. In addition, members can develop communicative and social abilities in seminars oriented to some extent specifically toward volunteer organizations.

Another challenge is presented by the starting point for public transport lobbying work described in Section 1.2.1. Between the dominance of private car traffic and the rather limited possibilities for interest representation in PT, it is doubtlessly easy to loose the motivation that is necessary to overcome this situation. Lobbying work in this field has to deal with this issue as well. This work can and should try to change policy makers' perceptions of the problem, but it usually requires perseverance and patience. This is all the more critical because this work is usually carried out based on pure interest in the topic and the corresponding idealism.

What can engagement in transport politics deliver? In the foreground is of course success in influencing political decisions – whether it be 'supporting' in providing additional means, or 'prohibitive' in dealing with service cuts. There are indeed other effects as well. Even if criticism is not successful in individual cases, it can sensitize both decision-makers and the public and thus affect the way that decisions are made in the future or effect more extensive discussion before decisions are made.[5]

Also, the possibility should not be ruled out that over the course of time a change in thinking might take place and original opinions be revised. In such cases it is difficult to claim success, but nonetheless, the goal is reached.

4.3 Participation in Planning

Christian Bodensteiner and Nancy Neugebauer[6]

4.3.1 Historical Outline of Participation

The term participation refers to various processes of cooperation and involvement between people. A universally applicable definition of the term is difficult to find, since various research disciplines such as sociology, political

5 For examples of such effects see Kamphausen (1999); Emmerson (2005).

6 Section 4.3.7.

science, economics, and pedagogy tend to use it in different ways. The term's definition therefore always depends heavily on context.

In the area of transport planning, participation means the '*involvement of the public in planning-related decision making processes*'.[7] During planning, the term participation is often equated with *civic participation*. In this case, participation can exhibit various degrees of intensity, ranging from the civic participation at the early stages of urban development planning to the integration of citizen initiatives and decisions.

In the West German political discourse of the 1970s and 1980s, the term participation was regarded a 'specifically political buzzword'.[8] Participation was the core element of the *democratization efforts* of the mid-1960s, which had demanded increased civil participation in exercising political power. These demands reflected a view that emerged during these times, which held that securing the common welfare was not merely the duty of the state, but rather that of every individual citizen.[9]

In addition to rather general democratic principles, the elements responsible for these demands for greater participation were primarily planning procedures in the area of infrastructure development, but also in urban development and redevelopment. Since the 1970s, projects and arrangements decided purely by politicians and administrators usually generated much more criticism than they had just a few years before. Controversial projects from those times like the nuclear power plant in Kalkar or the third runway – 'Startbahn West' – at Frankfurt/Main airport were often considered environmentally dangerous and the more critical populace – and in particular the citizen initiatives whose numbers had skyrocketed since the 1960s – fought against them. The resistance was primarily concerned with *planning procedures deemed undemocratic*. This criticism was affected by the politics of the 1970s, with the result that participatory elements were increasingly integrated into politics and administrative actions.

4.3.2 Formal and Informal Participation – Definition

Participation is formally regulated in many planning areas. For example, plans for expanding transport infrastructure are dealt with in planning approval procedures that legally mandate the involvement of a third party in the planning process. These could be other public functionaries who are tangential to the plans, but also interested individual citizens. A legally regulated form of participation such as this is referred to as *formal participation*.[10] Examples of such formal practices are public coordination during development planning and planning approval procedures, or – in the case of transport companies

7 Fürst (1998), p. 6.
8 Holst (2001), p. 5.
9 See Holst (2001), p. 5.
10 Böhme et al. (1998), p. 4.

– practices for customer involvement that are determined in the company's by-laws (see Section 4.4).

'Informal practices' go further than formally regulated agreements. In *informal participatory practices*, also called 'cooperative practices',[11] the range of topics and methods is usually larger. Informal practices tend to be more cooperative. They often entail working in groups and with methods designed for these group processes that contrast with the 'typical bureaucratic request-review-recommendation model'. In this sense, informal participation is still very structured and formalized. Usually processes in informal practices are designed to be open-ended. Some examples are citizens' exhibitions, brainstorming workshops for future activities, or planning cells. An extensive presentation of such practices in relation to public transport can be found in Section 4.5.

Formal and informal participatory practices can *overlap*. This grey area results when, for example, additional cooperative methods and practices are employed in the scope of formally regulated participatory practices. Passenger advisory boards could be considered a grey area. While passenger advisory boards are often established in transport companies' bylaws, the tasks they undertake are often freely elected and their method of working is often quite group-oriented (see Section 4.6.3).

Informal meetings of interested parties and political or administrative representatives are another form of participation. Here, the primary concern is the exchange of opinions in order to have an impact on decisions (Lobbying, see Section 4.2).

4.3.3 Levels of Participation

Theoretical treatises about participation usually assume there are various dimensions or levels of participation. Selle,[12] Lüttringhaus,[13] Klotz[14] and Brinkerhoff and Goldsmith[15] describe participation as a succession of levels leading from relatively passive to active modes of participation.

With every level, the intensity of the participants' involvement is increased. According to the description by Brinkerhoff and Goldsmith, the levels are: 'information sharing', 'consultation', 'collaboration', 'shared decision-making' and 'empowerment'. These levels can be represented in a participation pyramid (Figure 4.2).

With 'information sharing', information flows only in one direction. At this level, the initiator (usually the government) provides information about the intended plans. 'Consultation' implies a two-way flow of information as well

---

11 See Schröder (2003), pp. 2 et seq.

12 See Selle (2000), p. 62.

13 See Lüttringhaus, (2003), p. 66.

14 See Klotz (2002), pp. 15 et seq.

15 See Brinkerhoff and Goldsmith (2000), pp. 4 et seq.

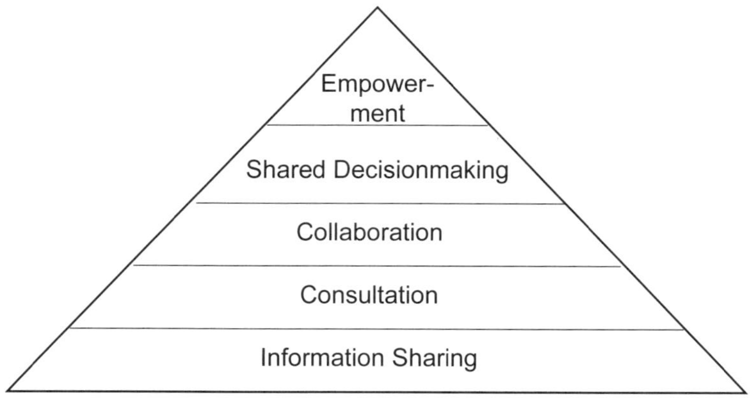

Figure 4.2 The participation pyramid

Source: Brinkerhoff and Goldsmith (2000), p. 5.

as an exchange of ideas between the initiator and other participants. At the level of 'collaboration', the initiator/process leader invites other groups and participants to collaborative consultations, while retaining the authority to make decisions. At the next level, the initiator surrenders this decision-making authority to other parties ('shared decision-making'), who then collectively agree on decisions. Participatory practices in which such collective decisions are made are called cooperative practices.[16] At the final level, 'empowerment', the ability to make decisions is completely surrenders to the participants.[17]

The EU project GUIDEMAPS uses a form of the typology of participatory practices that is reduced to two levels.[18] A distinction is made between 'information giving and gathering' and 'interactive engagement'. In the first case, the flow of information is one-sided, while 'interactive engagement' is a two-way flow of information.

4.3.4 Goals of Participation

Participation (from the government's point of view) pursues diverse and differing goals that vary widely from case to case. Nevertheless, specific goals that are sometimes closely related can be identified during the analysis of planning processes. Potential motives for participatory practices include: gained insight, early warning system, conflict avoidance, acceleration of the voting process, as well as increase to acceptance/legitimation of plans.[19]

16 See Selle (2000), p. 62.
17 See Brinkerhoff and Goldsmith (2000), p. 5.
18 See Kelly et al. (2005), pp. 64 et seq.
19 See Bodensteiner (2005), pp. 32 et seq.

1. *Gained insight*: an essential goal of civic participation is gained insight or information. This applies not only to products and services designed to meet potential customers' needs, but also for conducting planning. Through participation, planners can learn more about the topic or area and about the positions and possible attitudes of the people who live there. This is important because planners normally have a superficial knowledge of the area; that is they do not know it from personal experience.[20] In addition, experts and specialists usually regard their planning tasks from a specific perspective. Due to the job they have to perform, only certain portions of reality are of interest. The perspective these experts have differs strongly from that of those who are affected by and have to come to terms with the plans.

   Combining these two different viewpoints by means of civic participation reduces such deficiencies in perception and allows a more holistic view of the problem areas. This should help avoid planning errors.[21]
2. *Early warning system*: closely related to gained insight is the possibility, with the help of civic participation, of becoming sensitive to the areas that are particularly liable to develop problems in the course of the planning process. In this regard, civic participation can be seen as a kind of early warning system for further planning. 'If you provide information about intended plans early on, you will soon find out soon enough where any resistance lies.'[22] When this knowledge is acquired at an early planning stage, plans can still be easily altered accordingly, or other ways can be examined for minimizing a given protest. In this way, participation and early information gathering can act as an early-term resonance test.

   This is of particular importance if – in the case of local transport – users only have limited freedom of choice with regard to usage or non-usage (captives). Due to the largely monopolized structure of the PT sector, there is also frequently no choice between competing service providers.[23]
3. *Conflict avoidance*: it is often mentioned that difficulties and resistance can be avoided through civic participation. Resistances to planning can express themselves in various forms – the scope reaches from criticisms disseminated by the media to public criticisms by affected parties and citizens to the normal channels of judicial complaint and protest procedures. This resistance then leads to delays in the planning processes and possibly to decisions in favour of just one party, rather

20 See Fürst et al. (1998); Mettler-von Meibom 1999; Selle (2000), p. 169.
21 See Selle (2000), p. 43.
22 See Selle (2000), p. 170.
23 Konrad and Truffer (2003), pp. 2 et seq.

than making use of the available latitude for consensus. To avoid such blockages and restrictions, timely civic participation that brings together as many involved parties as possible and strives, in an informal manner, to achieve consensus amongst them can be quite effective.[24]

4. *Gains in legitimation and acceptance*: civic participation is conducted with the goal of increasing the legitimacy of decisions. For this reason, greater acceptance for the project during planning should be generated among planning officials and the population. An important goal of participatory practices is ultimately to increase the chances that plans will be accepted and executed. Civic participation should help to avoid the rejection or failure of plans.[25] Especially in politics and planning, such practices could improve the legitimation of decisions and operations.[26]
5. *Other goals*: in the literature, communicative practices are described as both a possible 'element and expression of a functioning democracy',[27] which can, by offering their support, enable even marginalized groups to take part in planning and politics.[28] Stimulating residents and (local) businesses takes advantage of existing know-how and increases the readiness for action.[29] Furthermore, communicative practices can result in greater self-confidence, greater personal responsibility, emancipation, and self-determination among the populace.

   Communicative practices are also seen as a possibility to (re-)establish trust in institutions among citizens as well as a means of generating satisfaction. Especially in politics and planning, such practices can increase the connection between specialized technical issues and citizens.[30]

   On the socio-political level, the concepts of social integration and the opportunity to help citizens help themselves both play a role. In addition, the implementation of (collectively developed) solutions can serve as a 'visible signal' that one has taken the interests of a given district into consideration in order to prevent further damage to one's image and further drifting apart between districts.[31]

24 See Selle (2000), p. 170.
25 See Faust (2001), p. 184 et seq.
26 See, for example, Barnes (1999), pp. 60–75; Webler and Renn (1995); Selle (2000).
27 Selle (2000); Barnes (1999).
28 See Schröder (2003), pp. 2 et seq.
29 Authority of Health and the Environment, Hamburg (2001).
30 Barnes (1999).
31 Authority of Health and the Environment (2001).

4.3.5 Ground Rules for Participatory Practices

General basic requirements and quality criteria for participatory practices can be formulated and applied to the implementation of participation models. The most important rules for participatory practices are:

Earliness The sooner participation happens, the better. It is especially important at the beginning of planning processes, as this is the time that fundamental decisions are made that are decisive in guiding future planning. If participation is lacking in this phase, substantial changes further along in the process are only possible with great difficulty. Also, questions about the process can be more easily clarified at an early stage.[32]

Transparency Transparency is an essential component of every participation process and a crucial element of its legitimation. Questions for participants as well as their tasks and roles must be clearly expressed and accepted right from the start. Also the goals of individual parties must be known to all participants. Transparency also means open information policies and conflict resolution in order to best handle misunderstandings and perceptual discrepancies. This necessitates that intermediate and end results are made accessible to all participants and to the public at large.[33]

Comprehensive participation Participatory practices should be comprehensive. The term 'comprehensive participation' implies that all important aspects and components of the planning process are sufficiently and extensively discussed in the course of participatory practices. This also requires the participation of a sufficiently large number of relevant participants.[34]

Implementation An essential prerequisite for participatory practices is the question of whether and how the results of the processes and discussions will be acted upon and implemented. If it is not certain that the political entities or other important parties will give the results serious consideration, it would not make any sense to conduct them. This kind of liability can be extremely frustrating for the participants.[35]

Openness of the results None of the planners should try to push through previously determined decisions as the result of participatory practices. All participants must be prepared to set aside their own preferences in the interest

32 See Selle (2000).
33 See Beckmann and Keck (1999).
34 See Bodensteiner (2005), pp. 32 et seq.
35 See Selle (2000), p. 46.

of finding the best solution for all parties involved. For this to happen, all participants in a given session must be treated as equals.[36]

Integration – stimulation of participants An important structural criterion for participatory practices is the integration of both organized and unorganized citizens and participants. Organized citizens, by which is meant primarily citizens' initiatives and associations, are easy to stimulate into participation, though they often pursue their own agendas. This is not so negative per se, though an event organizer or moderator seeking to attain as many opinions as possible should ensure that organized citizens are not the only ones contributing to discussion.[37]

So that a socially broad basic consensus can be attained during planning, intentional participation of all levels of society must be represented. Since it has been observed that primarily the more economically comfortable and better-educated classes actively take part in participatory practices,[38] stimulating the weaker levels of society is of crucial importance.

The above list of formal criteria describes important ground rules that are always to be observed in participatory practices. If these ground rules are not upheld, participatory practices will be subject to criticism. The risk that the process will fail is higher if the rules are not followed.[39]

4.3.6 Problems of Formal Participatory Practices

Formal participatory practices as legally prescribed in urban and traffic design exhibit a series of deficiencies in both their practical implementation and their approach.[40]

1. *Superficiality and minimal response*: experiences with formal forms of civic participation are not exactly encouraging. Only a minority of citizens attend the participation events. In addition, participation is often merely superficial.
2. *Often only organized interests take part:* there is a wide gap between the theory and the practice of organized civic participation. All too often, civic participation is restricted to the voicing of organized interests.
3. *Socially selective*: traditional political participation is unequally distributed in terms of social structure. Highly educated people, those with advanced career positions, middle-aged men and public servants

36 Beckmann and Keck (1999), pp. 20 et seq.
37 See Linder and Vatter (1995), p. 186
38 Lüttringhaus (2003), p. 68
39 Bodensteiner (2005), SS 31 et seq.
40 Taken from Reinert (2005)

tend to dominate. Foreign citizens, youths, women, older employees and people with lower income levels are underrepresented.

4. *Time problems*: even if politically interested, many citizens simply do not have the time to engage in such activities.
5. *Civic participation often takes place too late*: interest and engagement primarily result when one is personally affected. Decision processes are often too far along for anything but the most minimal influence to be made.

A main problem of civic participation is the difficulty of motivating the populace. Citizens have to be persuaded to take part in participatory practices. Interest and engagement result 'in the population only when one is personally affected, taking the form of concrete suggestions for measures or actions. But this is often only recognizable at a very late stage in the planning and realization process.[41]

Participatory practices in which changes are almost impossible due to an advanced stage in the process raise suspicions of being merely 'alibi events'. The risk is especially high when participatory practices occur late (cf. Figure 4.3). It is easy to come away with the impression that such participatory practices only aim to 'manufacture public acceptance for previously made decisions'.[42]

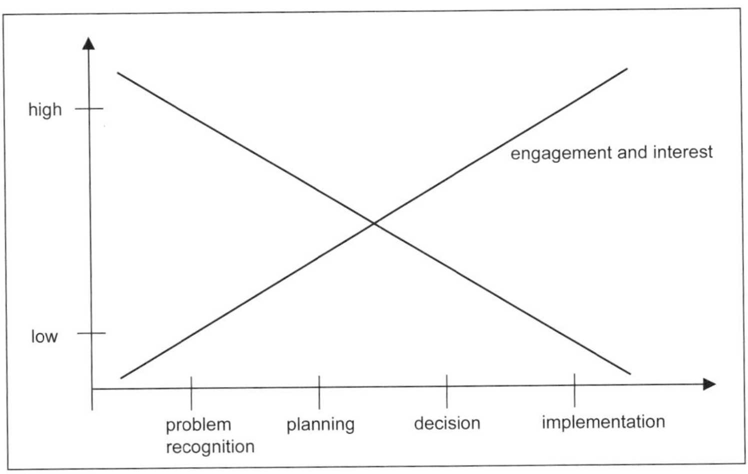

Figure 4.3 Increase to participants' interest and simultaneous decrease to the possibilities of exercising influence in the planning procedure

Source: author's depiction.

41 Böhme et al. (1998), pp. 5 et seq.
42 Faust (2001), p. 186.

6. *Tendency of segmentation*: in society there is a tendency to seal oneself off (segmentation). Different groups discuss more about each other than with each other. Confirmation of opinion rather than exchange of opinion is sought.

4.3.7 Good Organization of Group Processes

In the scope of participatory practices, many stakeholders who often have different views and interests come into contact with each other. Constructive collaboration of all participants is a crucial factor for a successful event. Requirements for the success of the process are:

- *thorough preparation* and structuring of the collaboration;
- fruitful *collaboration within the group* through good integration of all participants, sufficient flow of information, well-qualified coordinators (especially in the area of conflict management); and
- *comprehensive processing of the results* of the process.

Thorough preparation of collaboration In the starting phase of such an event, an analysis of the stakeholders should be conducted to identify key people (potential partners and people or interest groups that are to participate). In addition, goals should be set and the interests that are to be taken into consideration should be determined. In the end, a detailed process plan and schedule should be created.[43]

Collaboration in group processes Under what circumstances is effective collaboration possible? The productivity of a group comprised of diverse stakeholders and characters is dependent on many factors, which will presently be described further:

1. *Guidance of the group*: in the course of participatory practices, guidance is principally managed by the event moderators. In contrast to ordinary leadership functions, the moderator plays a supporting and accompanying role: s/he coordinates, initiates discussion, gathers feedback, takes everything into consideration and mediates any conflict that might arise. In addition, s/he has an 'icebreaker and motivational function':[44] It is her/his responsibility to open a creative space that encourages active participation and to keep the conversation going. On the one hand, the moderator plays a key role and her/his experience, intuition, and ability are decisive in determining the quality of the collaboration. On the other, situational factors are also relevant for

43 See Schmithals et al. (2004), pp. 132f et seq.
44 See Burow (2002), p. 78.

the moderation to be successful. Some examples include the structure of the tasks to be completed (for example, degree of complexity), the participants of the group process itself (for example, qualifications, experience), and the group (for example, size, cohesion).[45]

2. *Composition of the group*: participatory practices involve the interaction of people with very different qualification and experiences. Heterogeneous groups comprised of individual whose sex, qualifications, or experience differ often work with more openness, flexibility, creativity, and effectiveness. At the same time, such heterogeneity within a group raises the risk of conflicts, minimizes internal cohesion, and makes reaching a mutual consensus more difficult. In addition to technical and contextual qualifications, social skills and the ability to work as a team are decisive factors.

   The size of the group is also critical for the success of the process. On the one hand, a group's performance decreases as its size increases; on the other hand, a group having too few participants is easily overwhelmed, which also lowers overall performance. For processing specialized, complex issues, a group of six to eight experts is recommended. Less-specialized tasks can be handled by a group of up to 30 people.

   The duration of a group's collaboration can also have an effect on its performance. If a core group of members works together over a long period of time (for example, passenger advisory board), there is a risk of establishing rigid structures and losing openness to outside viewpoints. At the same time, having a common history, and a series of challenges and conflicts that were overcome together can be quite conducive to a group's cohesion.

   Given the concerns detailed above, it is important to take into consideration the tasks to be completed, the projected (work) duration, the size of the group, and the requirements with respect to qualification and suitability of members.[46]

3. *Exchange within the group*: first it must be ensured that all participants are fully informed about the participatory practice at an early stage. This applies to information regarding the content of the decision at hand as well as to the participatory practice itself. In discussions, it is important to consider the concerns and opinions of all participants. It is the moderator's job to ascertain the various interests and topics.[47]

   In addition, successful communication between the participants is a central factor for successful collaboration. For a constructive understanding, the following principles can be used for orientation:

45 See Böhm (2006), pp. 49 et seq.

46 For more details see Böhm (2006), pp. 36 et seq.

47 See Fischer et al. (2003), pp. 76, 84.

- communicate constructively with each other,
- give and receive feedback,
- tackle difficult topics, and
- learn by balancing.[48]

It is also helpful to visualize important viewpoints, (intermediate) results, and decisions (for example using a board, flipchart, or projector). This contributes to transparency of the process and the discussion 'because otherwise the process soon begins to spin its wheels'.[49]

Sufficient opportunities for unofficial communication between participants – for example, long breaks or a nightly outing – are beneficial to the process and help create a positive atmosphere at all meetings and events.[50]

4. *Conflict management*: an unsatisfactory distribution of resources, disagreement regarding responsibilities or power, content-related controversies, and animosity or competition can cause conflict in the group process. These pose the risk of losing joy and engagement in the collaborative process, resulting in stagnation and resignation for the remainder of the process, ultimately leading to its failure. The challenge is to confront these tensions assertively, that is by initiating discussion of potential causes of conflict and not avoiding mention of unmet expectations in order to develop ideas for conflict solutions. Collective conflict resolution in which participants approach one another can strengthen the group's internal cohesion.[51]
5. *Post-processing*: last but not least, comprehensive post-processing of the practice conducted can help to improve the efficiency of future projects. A documentation of the process and its individual production steps and their results offers the opportunity to learn in retrospect.[52] At the same time, important realizations and results can be made available to the public in order to create an impact outside the group of participants and to generate attention and acceptance for the project.[53]

In closing, the necessary factors for collaboration can be summarized with the keywords shown in Box 4.1.[54]

It can be concluded that, of the multitude of relevant factors, thorough preparation, professional execution, and careful postprocessing are the most important for successful group work. In principle, one should always be

48 For more details see Blanckenburg et al. (2005), pp. 145 et seq.
49 Schäfer (2004), p. 107; see also Fischer et al. (2003), p. 85.
50 See Schäfer (2004), p. 112.
51 For more detail see Blanckenburg et al. (2005), pp. 189 et seq.
52 See Schmithals et al. (2004), p. 133.
53 See Burow et al. (2002), p. 10.
54 Author's depiction based on Böhm (2006), p. 68.

Box 4.1

- Collective, clearly defined, overall goals.
- Qualified moderator(s).
- Sufficient material, personal, and time resources.
- Autonomy.
- Internal communication, open exchange, transparency.
- Good personal relationships.
- Technical and social competence of members.
- Clear rules and standards.
- Geographic proximity/accessibility.
- Common vocabulary/language.
- Incentives.

aware of potential problems or hurdles that may arise in the course of such participatory practices in order to promptly confront and resolve deficiencies (see also Section 4.3.6).

4.4 Practices for Formal Citizen Participation in Public Transport[55]

Service planning in public transport is not the task of transport companies alone; it takes place in an exchange between political decision-makers, other sectors of planning, regulatory authorities, and other bodies. Even though planning laws and procedures differ in individual countries in terms of their details, planning and operation are indeed comprehensively regulated, and the steps that must be followed are extensively predetermined. Important tools in this process include:

- (strategic) plans for developing transport infrastructure and services;
- contracts between companies and public transport authorities about the extent of the service, quality standards, etc.; as well as
- qualitative requirements for the company based on legal obligations, concessions, and so on.

Such practices are highly binding. They are thus of great importance to the question of whether and how the public – including passengers – are allowed to have insight into and exercise influence on planning processes. If there is participation provided, this generally guarantees a minimum of information and commentary possibilities, even if the purpose and type of participation (for example formal parameters) are often narrowly limited.

55 For authors' names see section headings below.

However, there are two considerations in this process: firstly, a formally secured requirement for citizen participation can be interpreted and implemented in very different ways, especially if – as is definitely common – the requirement only states that there must be a form of participation, leaving the type and method of participation to the respective local bodies. This leaves room for very comprehensive, and for very parsimonious solutions.

Also, it is often the case (see the national studies in Chapter 5) that there is no participation explicitly provided for representatives of the public and/or users in PT planning procedures. In this respect, the practices presented in the following sections represent positive exceptions rather than contemporary practice, even if many of them have a long history of application to look back upon.

4.4.1 Possibilities for Public Influence in the Framing of Transport Contracts – A Comparison of British and German Approaches

Uwe Böhme

In local public transport in Germany, especially in local passenger rail transport (LPRT) an increasing number of transport services are being turned over to the competitive market. In 2002, c. 11 per cent of LPRT transport services were transferred to the open market.[56] In a competitive market, the awarding of a contract requires an agreement between a transport authority and a transport company.

In LPRT, the states decide how their local transport services should be designed. In order to implement these specifications and to purchase transport services, states mostly form public transport authorities, which act on behalf of the respective states. These authorities, along with the transport companies, regulate the framing of contracts for transport services purchased by the state. For this sort of award procedure, there is a question of how much and at what stage public influence is possible, and of how it should be regulated. The award process can be divided into the phases represented in Figure 4.4.

Possibilities for influence in the preparation phase The structuring of contract awards in LPRT is done by the transport authority alone. There is a procedural method determined by both the official contracting terms for the award of public service contracts and the Public Procurement Code (PPC).[57] There is no provision for public participation in, for example, the framing of the call for tenders.

However, informal participation is possible in the work proceeding the award of contracts. Passenger advisory boards have been formed in some states

56 Wachinger (2006) p. 42.

57 Framework regulation for public service contracts, 'Verdingungsordnung für Leistungen' (VOL/A).

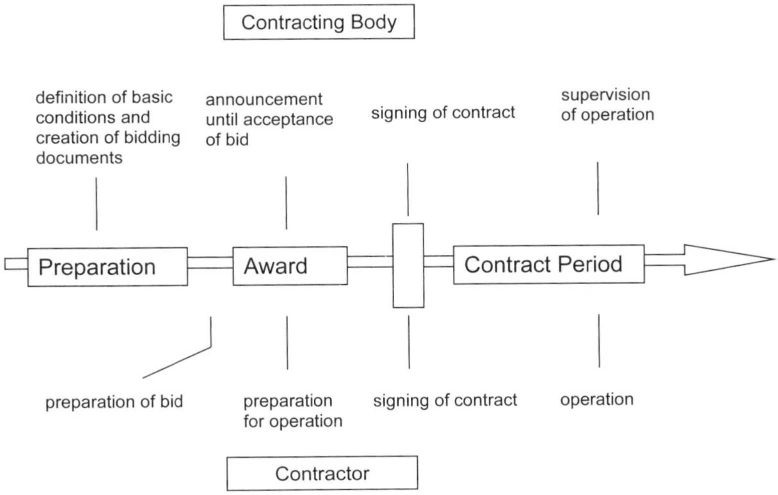

Figure 4.4 Structure of the contract awarding process

Source: author's depiction.

during the course of regionalization; these can be used to gather information about customer expectations (see Section 4.6.3). For example, a LPRT advisory board was formed by the Verkehrsverbund Berlin-Brandenburg, (Berlin-Brandenburg Transport Association) (VBB) that consists of 20 members including organizations such as the Verkehrsclub Deutschland (German Transport Club) (VCD), the Interessengemeinschaft Eisenbahn, Nahverkehr und Fahrgastbelange Berlin (Railway, Local Transport, and Passenger Interest Syndicate Berlin) (IGEB) and the Allgemeine Deutsche Fahrradclub (German Public Bicycling Club) (ADFC). The advisory board is 'heard' and may express its opinions about such things as accessibility or bicycle carriage.

There are no other possibilities established in the aforementioned regulations and statutes for individual citizens to exercise influence in the preliminary stages of awarding a contract.

In some individual cases participation can, however, be more intensive. For example, in Schleswig-Holstein LPRT there have been special route advisory boards established, which provide for close contact to institutions involved in the decision-making process. It can thus be concluded that the intensity and effectiveness of advisory boards' participation in Germany depends on the respective region as well as the organization of the institution that is offering the bid, and often takes on the character of *informal briefing*.

Possibilities for influence in the contract-awarding phase There are very detailed requirements for the awarding of public service contracts, which are determined in the official contracting terms as well as the Public Procurement Code (PPC). These documents also include specifications for the appraisal of contracts. For reasons of confidentiality and neutrality, it is required that only a very small circle of individuals are allowed to see and appraise transport companies' offers. Neither the official contracting terms nor the PPC provide for any kind of public participation. Thus for formal/legal reasons, there is *no possibility for public participation* in the process of awarding contracts.

Possibilities for influence during the contract period (operation period) The main activities of passenger groups and formal advisory boards occur in political lobbying work during the time that transport services are being delivered. This phase first introduces the capabilities and deficiencies of readjusting transport services, to which transport associations and other organizations can draw attention. This happens mostly through publicity in appropriate media such as periodicals and trade journals, or newspapers as well.

However, the experiences of train user associations have shown that changes do not necessarily have to occur through public pressure alone. Often, cooperation takes place between train user organizations, transport companies, and transport authorities on an informal level, wherein ideas and differences of opinion are exchanged. Sometimes it is possible to achieve faster and more effective results in this way than it is through formal practices.

Transport contracts that have been made between states and transport companies are generally not available for public viewing. There are thus no exact official reports available about basic financial parameters or agreements about things such as punctuality and according malus payments.[58]

It is thus difficult for citizen and user groups to exercise influence. The fact that a significant amount of tax money is used for LPRT is consistently used as a reason for making these contracts public.

It can ultimately be said that *public participation* in the awarding of transport contracts is fundamentally *not possible* in Germany. In some states there are however advisory boards, which are given the possibility to express their concerns in the preliminary phases. The next possibility for exerting influences occurs in the time between the signing of the contract and the start of operation.

In contrast, there are more possibilities to exercise influence for rail passenger associations in *Great Britain* where contract awards are done primarily through the Department for Transport (DfT). The Department is required to hold a consultation with interested districts and organizations *before issuing an award.* A so-called consultation document is thus produced in the preliminary stages of the process, which states benchmark figures for

58 For an example of a transport contract with a bonus/malus system see Section 3.5.1.4).

the contract awarding process. In addition, a passenger satisfaction analysis is carried out for the routes involved. Among the organizations involved is the state-recognized Passenger Focus. Further possibilities for exercising influence on the actual awarding of contracts, however, do not exist in Britain either.

In the actual *contract-awarding phase*, however, Passenger Focus has no direct possibility to exercise influence. It is especially not intended that Passenger Focus take part in the contract negotiations.

During the *operation period (contract period)* transport companies are legally required to announce their punctuality figures to the DfT as well as to Passenger Focus (see Section 3.5.2.1). Furthermore, Passenger Focus must be informed of route shut-downs as well

Unrelated to these functions, Passenger Focus also has the possibility to take part in the decision-making process and in the monitoring of service delivery through lobbying work during the operation period.

4.4.2 The Swiss Timetable Procedure
Martin Schiefelbusch

In Switzerland, the so-called Timetable Procedure (Fahrplanverfahren) has long been implemented and legally supported, providing public participation in a form that, compared to other countries, is quite extensive. The timetable procedure has its foundations in Article 7 of the Timetable Ordinance, which regulates the production and publication of timetables as well as operational requirements. It applies to all forms of public transport with the exception of lines limited to one locality only. The procedure's fundamental steps are summarized in Figure 4.5.

The foundation for citizen participation in the process of timetable design is Article 7 paragraph 1 of the Ordinance, which however only contains a very generally stated requirement for this: 'During the course of the timetable procedure, the cantons [regions] will hear the interested groups in a suitable manner.' Its implementation is up to the cantons; after the conclusion of the hearing, they make the decision about how the matters at hand should be handled. The Bundesamt für Verkehr (Federal Department for Transport (BAV)) provides for the dissemination of the plans – and ultimately for their publication on the Internet, and coordinates the procedure on the national level, but usually does not comment on the contents of the plans.

In this respect, it is inaccurate to refer to 'the' timetable procedure; the following descriptions thus do not claim to reflect the regulations in all of their detail. They refer essentially to the experiences of the procedure in the year 2004. Alongside experiences on the national level, the procedure in the canton Thurgau will be described.

Conduct of the timetable procedure For a long time, the publication of timetable plans took place through public presentation within the canton

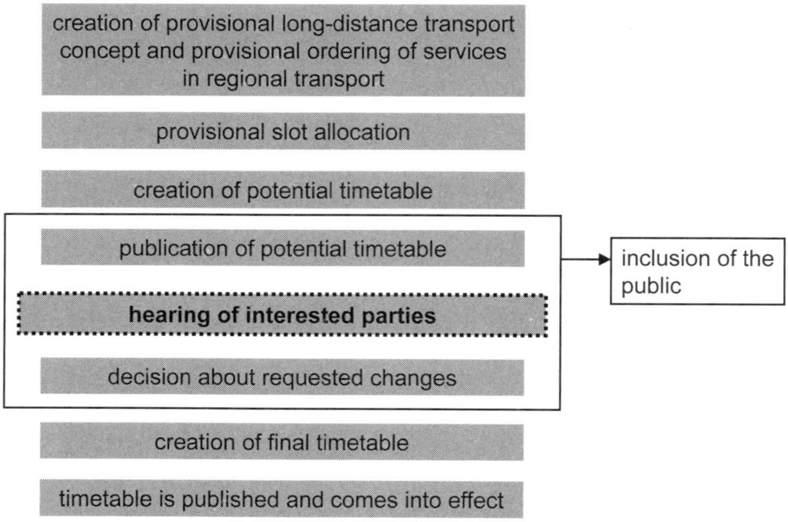

Figure 4.5 Fundamental steps of the timetable procedure

Source: author's depiction.

and municipal administrations. It is thus comparable with the presentation of plans for infrastructural measures in the context of the plan-approval procedure. In the timetable procedure of 2004, the plans were for the first time additionally publicized on an Internet site (www.fahrplanentwurf.ch). This publication on the Internet met with great interest, also because of the extensive nationwide changes in the 2005 timetable created by the launch of the newly developed route Mattstetten–Rothrist. During the presentation time, a total of 1.7 million visits by 259,100 people were registered, which lead to 6,300 submissions over the Internet.[59] An additional c. 7,000 submissions nationwide were sent directly to the cantons.

From the customer's perspective, the possibility to express an opinion about timetable plans is an essential element of the timetable procedure and the basis for the following steps. This first step, however, still does not say anything about the handling of comments, this is decided only in the following consideration and discussion.

The design of these elements shows the essential differences between the individual cantons, along with the differing relationship to further planning steps. The 'minimal form' of the participation thus consists of the discussion of the submissions alone between the cantonal department and the transport companies, the most comprehensive form in a public discussion, to which not only objectors but the entire public has access. The publication of the

59 BAV (2004).

decisions is designed in different ways as well. In most cases to date, there has been no explicit reply to objectors, instead the result was available to be seen only when it was published in the final timetable. The results of the content review are, however, mostly visible through, for instance, publication in official journals and similar texts.

The schedule and the further steps are presented in Table 4.1. This makes it clear that there are only around four weeks available for the content review of the comments, discussion with all parties involved and decisions about possible changes.

Table 4.1 Schedule for the timetable procedure

Stage	Date
Orderer's specifications:	
allocation of funds to the cantons, long-distance transport concept	22.8.03
advance notice of ordering in regional transport	19.9.03
Preparation of bids by the transport companies	5./24.12.03
Contract negotiations	Until 5.3.04
Preliminary route allotment	17.3.04
Publication of potential timetable	16.4.04
Deadline for submissions	10.5.04
Submitting requests for changes by the cantons	11.6.04
Definite ordering and slot allocation	13.8.04
Publication of timetable	11.10.04
Timetable coming into effect	12.12.04

Source: BAV.

For their implementation, the transport company then has another ca. two months. This already shows the two essential problems of the procedure:[60]

1. the participation takes place under significant time pressure; and
2. in a late stage of the service planning.

Thus only small changes to the forthcoming timetable can be settled in the context of the timetable procedure. The public is often not conscious of this and so have many false expectations about the effect of a submission. The timetable procedure is nevertheless used to cultivate fundamental criticism and not only to generate discussion on fixed points. If this is not considered,

60 See Michel (2000).

it is easy to get the impression of an 'alibi event' in which citizens' opinions in reality play no role.

Indeed, it is current practice to use submissions that were not considered due to time constraints as follow-up for the preparation of the next timetable. This is, however, not always satisfactory in the view of objectors. This is unfortunate from the planning perspective as well – if truly serious deficiencies are to be discovered or new ideas developed during the course of the participation, this should happen at a point when changes can still be made.

Swiss planners and passenger associations are also aware of these difficulties, and the result is that what was previously described as an 'official' planning procedure is now seen as a corrective instrument. As such it certainly is qualified – by the principle that four eyes see more than two – to review the plans in terms of mistakes and discrepancies that professional bodies might miss due to 'routine tunnel vision'. Further steps are necessary to establish a true possibility for co-design of services. Thus, some of the cantonal procedures have begun to provide for more participatory elements.

A model for other countries? Citizen and customer participation in service planning takes place on a voluntary basis in other countries as well. These procedures could provide significant stimulation for planning as well. As experience with the procedures shows, participation at early stages and 'supervision' of timetable designs should compliment rather than replace one another. Based on the Swiss example, some German transport authorities have recently started to publish timetable designs – generally only on the Internet.[61] In other countries, this type of consultation seems to be unknown, although consultation with user representatives on service planning in general is more common.

Particular advantages of the Swiss procedure are:

1. the function of the timetable procedure as a mutual crosscheck which enables timely discovery of factual and editorial mistakes;
2. its value as a corrective medium between the preparations of local transport plans and transport contracts;
3. the obligation to deal with submissions from and give feedback to all parties involved, which is a welcome contribution to strengthening PT in the community and building acceptance.

Engaging with submissions and comments doubtlessly requires – as does every form of market research – the allotment of sufficient resources. It should, however, be remembered that timetable data, at least in LPRT, have to

61 Such as the Nordhessischer Verkehrsverbund (North Hessian Transport Association), and the Nahverkehrsgesellschaft Sachsen Anhalt (Transport Authority of Saxony-Anhalt).

be determined at least eight months in advance because they are necessary for the ordering of railway lines.[62] The extent of the demand to be expected could essentially depend on the success of actually channelling comments that refer to other steps in the planning process to those points in the process. In other words, the better the organization of participation and handling of customer expectations in the preliminary phase, the fewer complaints are expected in the detail phase.

4.4.3 The Berliner Verkehrsbetriebe's Customer Dialogue Process *Christian Bodensteiner*

Since 2004, the Berliner Verkehrsbetriebe, 'Berlin Transport Services' (BVG), has employed a special form of customer participation. With the changes of timetables that occurred in December 2004, the BVG changed a large portion of Berlin's bus and street tram network. In the preliminary phases of the so-called Metrolinien-Plan (metro route planning), the BVG carried out a first-time citizen participatory practice. The core of this practice, called 'customer dialogue', was the possibility to consult about alternative routing options in the early phases of the planning process.[63]

The BVG instituted a public vote to decide routing plan alternatives for the bus and/or street tram networks in 14 zones of the Berlin metropolitan area. For each zone, there were two possible variations presented, and citizens were given the opportunity to vote for their preference (see Figures 4.6 and 4.7).

The public could take part in the form of street polls/information stands as well as through online polls.

Alongside the vote about alternative routing options, this citizen participatory practice also offered the opportunity to express *qualitative* ideas about the LPT network. This gave the BVG the possibility of hearing other kinds of opinions, suggestions, and criticisms as well, and in some cases to incorporate these into the planning process.

Over the course of this seven-week process, over 24,000 citizens participated. Among the items, there were very clear decisions for some options, as well as votes that were very close or undecided for others. The *results* of the vote were largely implemented.

In addition, the qualitative inquiries allowed the BVG to hear a wide array of citizens' *suggestions* and to integrate them into the routing plans at an early stage.

Because the design of timetables in public transport generally does not provide for citizen participation, it is very positive that a transport company would seek out a customer dialogue of this kind.

62 See Anon (2004).

63 Bodensteiner (2005), pp. 56 ff.

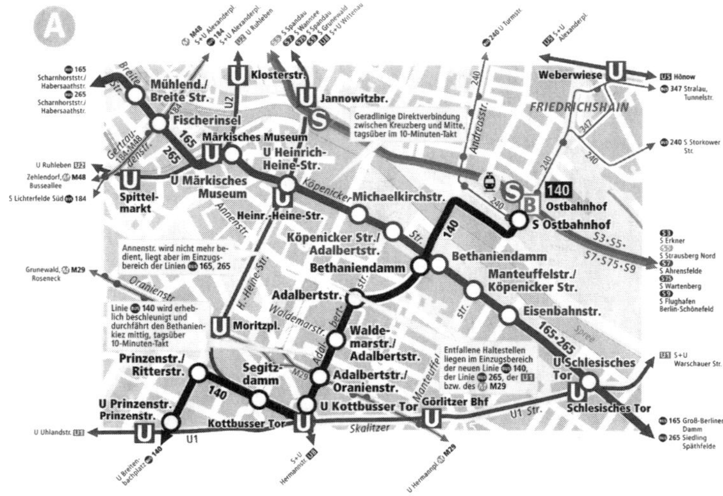

Figure 4.6 **Option A from a BVG ballot. This option offers two bus routes with fewer stops in the area, but an all-day service on both routes**

Source: BVG.

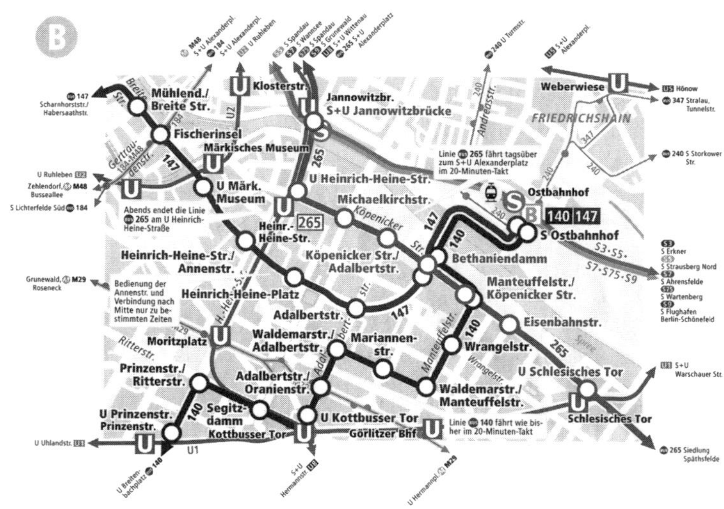

Figure 4.7 **Option B from a BVG ballot. This option offers three routes in the area and a better geographical coverage, but some of the services would not operate all day. The ballot asks, 'Which of these variations do you prefer?'**

Source: BVG.

Due to the strong public criticism of the BVG's Metrokonzept, the *evaluation* of this process notes a deficiency in the fact that planned cut-backs were not highlighted in the early stages of participation. Such a compartmentalization of issues should be critically questioned, especially in light of the participatory goal of facilitating legitimation.

The renewed implementation of a customer dialogue'using a ballot with variants for the following timetable change has shown, however, that the company sees this practice as essentially positive.

4.5 Informal Participatory Practices

Jenny Schmithals

4.5.1 Stakeholders and Structures in Informal Participation

In order to develop sustainable transport planning, one that incorporates a social dimension by taking participatory aspects into account, it is necessary to change the common understanding of planning. Particularly in cases of especially complex and/or conflict-laden planning issues, adequate participatory practices should be provided in every phase of the transport planning process. These practices should be flexible enough to accommodate the entire planning process, according to the nature of its particular outcome. This cannot be achieved by formal participatory practices alone, which are provided by legal foundations or transport company statutes. The instruments applied by transport companies to date, for example, in the areas of market research, public relations, and complaint management have not adequately grasped users' needs. Thus, new forms of communication seem necessary.[64]

The criterion 'informal' does not refer to a lack of structure in the participation process; on the contrary, there are various common elements that play a role in the process.

- *Information*: participants are informed as comprehensively as possible about the particular planning problem and exchange their views.
- *Communication*: misunderstandings are resolved and – in cases of controversial views – consensus- and decision-building is moderated.
- *Documentation*: the various interests and points of view are compiled.

In principle, the initiative for a participatory practice can come from any interest group. At the beginning, the initiators should conduct a stakeholder analysis in order to make sure that all people and/or organizations that would possibly be interested are included. A schedule should also be created with

64 See, for example, Böhme et al. (1998); Schiefelbusch (2005).

regard to dates such as school vacations and local elections so that stakeholders are not excluded or results skewed because of conflicting dates.

In choosing a participatory practice, considerations include not only the type of problem, but financial parameters as well. Is the local administration prepared to fund the project? Will developers perhaps invest in cooperative planning processes?

It is important that there is a solid agreement from the start about how binding the results will be.

> In all efforts at cooperation, legislative boundaries and legal jurisdictions, which are closely connected to a certain understanding of democracy, should not be overlooked. However, participation should not become a farce either. If classic institutionalized participation forms are to be expanded, fields of activity and the scope of decision-making processes must be delimited.[65]

Generally, the results of informal participatory practices must be conveyed into decisions through arbitration. It is thus advantageous for decision makers to commit themselves to finding a place in the planning process for the recommendations that are developed.

Potential stakeholders in the transport planning process are:

- transport users (for example, particular user groups);
- interest groups, initiative groups, and similar organizations;
- members of the administration (for example, representatives of country or municipal councils and/or district representatives);
- transport companies;
- external specialists.

The SRL[66] research group Beteiligungsverfahren (Participatory Practices Study Group)[67] makes a distinction between representatives of the institutional, the organized, and the general public. While the former are traditionally involved in planning processes, the stakeholders of the organized and the general public must be included through cooperative participatory practices.

The *results* of a consensus-oriented participation process can take the form of recommendations to policy-makers. It can be expected that participants' understanding of the planning process will be deepened. Citizens are valued and taken seriously in participatory planning processes. This can potentially

65 Böhme et al. (1998), p. 30.

66 Association for Town, Regional, and Country Planning (Vereinigung für Stadt-, Regional- und Landesplanung).

67 See Böhme et al. (1998).

contribute to their finding a new understanding of democracy. Disenchantment with politics can be reduced.[68]

4.5.2 The Potential of Informal Practices

Formal participatory practices, for example, those with statutory foundations or legal mandates, are often provided in a comparatively late stage of the planning process, if at all. This is often already past the stage when criticism that might have arisen about the planned project could have been addressed in the planning process and thus contributed to the project's acceptance. For example, residents' local knowledge is often left out of the planning process, and possible opposition thus left unidentified (see Section 4.3.6). In contrast, informal methods of citizen participation offer the following *opportunities*:

- the expert knowledge of users and/or residents, who gain their expertise through their everyday living environment, can be integrated in the planning process. 'The precise knowledge of a place and the everyday experience of various affected groups and citizen initiatives comprise a significant planning potential that the official planning process truly should not relinquish':[69]
- interests can be articulated and opposition identified early on in the process;
- misunderstandings can be resolved;
- cooperative planning processes have a trust-building effect and create mutual understanding;
- acceptance of planned projects generally increases if those affected are included in decision-making processes. The likelihood that planned projects will be successful is thus generally higher;
- possible conflict potential can be reduced, which can lead to an expedition of the process or to qualitatively more valuable solutions as well;
- participation can provide a contribution to equal opportunity and integration, especially for people who would not otherwise be accustomed to mobilising for their own interests;
- participation is closely connected to sustainable development. Participatory practices can contribute to the integration of ecological, economic, and social features of transport planning.

An adequate information policy is still not a participatory practice – it must provide *participation in decisions*. 'If citizens are to be taken seriously,

68 See Dienel (2002a).
69 Coffey (2002), p. 105.

it must be accepted that projects may be realized completely differently than planned – or not at all.'[70]

4.5.3 Potential Obstacles

Political decision-makers may potentially see participatory practices as threatening to their decision-making power, or they fear that processes will be slowed. Members of the administration may fear an (unnecessary) additional expenditure of time as well as money. Developers may think that participatory practices contain a risk for the realization of their projects, if the cooperative planning process is not committed to a specific result. Interest groups such as environmental organizations or citizens' initiatives may doubt the openness of the results and feel that they are in the weaker position. The citizens have the least professional mode of operation for the most part, and are possibly alienated and/or fear, similarly to the organized public, that they have little influence.[71]

4.5.4 Examples of Applicable Participatory Practices

This section provides an overview of significant practices. Some models that have already been carried out with great success will be described in detail in the preceding section.

Citizens' Assembly Developers, planners, members of the administration, and interested members of the public (for example residents or initiative groups) are invited to an information and discussion event. A large number of participants is desired. The local problems and planning project at hand are discussed. The event provides transparency and serves to analyse the general temperament.

Field trip/inspection The public is invited to develop a clear picture of the concrete planning process through a visitation of exemplary projects or a tour thorough the district. Proposals for the planning process can also be formed. The number of participants is limited.

Future workshop This method was developed in the 1970s by Robert Jungk and Norbert R. Müllert.[72] It helps in analysing a specific problem and developing visions for the future. The Future Workshop occurs in specific phases. Before the implementation phase there is first a criticism phase and a fantasy phase. A group of about 25 people including the general as well as the organized and

70 Coffey (2002), p. 104.
71 See for example Böhme et al. (1998).
72 See Jungk and Müllert (1989).

the institutional public works with this method from two to several days under the guidance of a moderator. Techniques such as brainstorming, fantasy trips, and creative games are combined. It is characteristic of this method that it does not start out by seeking realizable proposals; instead it first tries to develop utopias. The solutions created are so innovative that possible obstacles are more easily overcome.

Future Search Conference The Future Search Conference was developed by Marvin Weisbord and allows for a group of about 30 to 70 participants. The work takes place in two to three consecutive days primarily in small groups, and the moderators mainly stay in the background. It is their role to support the presentation of group work in the form of a plenum. The conference includes several phases: First, members reflect upon the past, then they analyse the present situation, and final develop visions for the future. In the end, members are meant to agree upon a common ideal for the future and develop measures for its implementation.

The experiences that have been gained through implementing this method have shown a strong mobilizing effect. This practice is well suited for incorporating specialists and interest groups who can contribute their respective perspectives with their input.[73]

Advisory Council Examples are passengers' or women's advisory councils. They often meet in the form of a 'round table', with a discussion leader and evaluate existing planning proposals. They try to arrive at a consensus-oriented solution and to develop recommendations.

Planning cell/Citizens' Report In a planning cell, ideally 25 randomly selected individuals work for a limited amount of time (generally four days) as citizen reporters in order to develop proposals for a specific planning problem. Members are compensated for not being able to attend to their professional or private obligations. A planning cell is guided by two moderators who develop a program that alternates between blocks with various specialist lectures and discussion rounds in small groups and in a plenum. The results are integrated into a Citizens' Report and made public.

Central features of this practice, which was developed in the 1970s by Peter C. Dienel, are random selection of participants, releasing participants from their obligations and compensating them for their time, as well as the program structure, process facilitation, and support by specialists. It is an advantage of the planning cell instrument that there is a very high probability of neutral recommendations that are oriented toward the common welfare.[74] The planning cell method has been applied frequently, especially for municipal

73 See Weisbord and Janoff (1995); Wachholz (2001).

74 Dienel (2002b); Stiftung Mitarbeit (1996).

issues. Citizens have been predominantly highly motivated and have also been in the position to familiarize themselves with the issues in the short time provided and to collaborate on an evaluation. Planning cells obtain an exceedingly diverse group of participants and have demonstrably influenced political decisions.[75]

Consensus Conference The form of the consensus conference has similar features to the planning cell. A jury of approximately 15 lay people are informed in depth about an issue through expert consultations. The consultation with specialists is public. Then, in a conference phase, the board of lay people give responses. In this method as well, the participants are not drawn exclusively from directly affected individuals. In contrast to the planning cell, there is no provision for compensation and no random selection, and thus the people who take part tend to be individuals who are already significantly more interested and engaged.

Mediation The practice of mediation is applied especially in cases of conflict. It can integrate a large number of participants (up to 100 people), and the process is lead by a neutral moderator. The proceedings are not public and are meant to include participants from as many relevant stakeholder groups as possible.

Citizens' Exhibition The Citizens' Exhibition method is applied in combination with the activating inquiry technique. In personal interviews, stakeholders are asked about their attitudes and goals concerning the issue at hand. Characteristic excerpts from these talks are combined with photographs to create posters as visual accompaniment (portraits of the interviewees, and/or pictures that incorporate the affected area/the living environment – that is a district). The exhibition opening has a decisive significance: all of the groups involved and members of the public who are interested in the issue are invited, and the event is meant to bring them into dialogue.[76]

This participatory practice is meant to give citizens affected by a planning issue a forum where they can express their desires and/or recommendations. Citizens' Exhibitions can be applied as an instrument and catalyst for creating a process of cooperative reflection toward participatory solutions to social problems. The greatest strength of the concept is the aesthetic and emotional power of the pictures and the quotations that go with them. It gives one a straightforward way of seeing citizens' perspectives, where statements are concretely connected with people and places. The Citizens' Exhibition thus takes on a strong activating function for the public. On the one hand, it has the internal effect of giving the participants a way to meet each other and

75 Dienel (2002b).

76 See Schophaus and Dienel (2003b).

exchange ideas. On the other han, the external effect is the opportunity for the public to inform themselves about the inner views of the subjects portrayed. In addition, inviting political representatives and the press creates a wide awareness for the subjects' interests and views. The exhibition opening creates a literal lobby for citizens who would otherwise have no access to the world of lobbying.

Transport forum A transport forum is oriented toward a group of participants that is as large and open as possible: The list of those invited includes all interested citizens, representatives of interest groups, as well as members of the administration, political decision-makers, and public transport authorities. The integrated approach employs moderated discussions at a 'round table' and field trips/on-site inspections. Planning proposals created on the administrative level are discussed and evaluated in the forum, and recommendations for improvement are developed.

Mobilization through evaluation The method of mobilizing evaluation was developed according to the practice of activating inquiry.[77] Affected citizens are asked about their opinions, and at the same time are motivated to mobilize for their interests and collaboratively create proposals for solutions, thus inciting a process of reflection through the questioning. Mobilizing evaluation 'thus prepares people to work with their own awareness of the situation and everyday problems, instead of imposing one's own concepts'.[78] The interviewer makes it his or her task to help the interviewees articulate their own positions.[79] The results of the inquiries are made available to all participants (the interviewees and everyone involved professionally in the planning process). Ideally, this communication leads to the opinions and ideas for solutions that arise from the process to be integrated in the planning process – there is no requirement for this to happen.

4.5.5 Case Examples of Successfully Applied Practices

4.5.5.1 The ÜSTRA Citizens' Report The planning cell method was used in local public transport for the first time with the Citizens' Report ÜSTRA. The method was selected to consider increasing quality and flexibility requirements for public transport. 'In planning cells people work together to find common solutions – younger people with older people, regular bus and train passengers together with notorious motorists, conservationists with members of

77 See Reinert (2003); Hinte and Karas (2003); Lüttringhaus and Richers (2003).

78 Seippel (2003).

79 Szynka (2003).

automobile clubs. All sides can only win from this exchange.' [80] The focus of the project was the current and future requirements for a customer-oriented transport company and for an attractive LPT. [81]

The planning cells, which were initiated by the PT provider ÜSTRA in Hanover, took place in 1995 and 1996 and were carried out in cooperation with the Stiftung Mitarbeit (Foundation for Cooperation). There were a total of 12 planning cells with 25 participants each, and issues addressed included the attractiveness of local public transport in Hanover, evaluation of travel time and speed, as well as safety, fare systems and citizen participation in LPT. The total programme in which these issues were addressed, comprised 16 work units. In each one, the citizen reporters were supported by a two-person moderator team and were advised by specialists.

The results included the recommendation that transport companies should value environmentally-oriented technology, and should offer a transparent fare structure, personal customer care, and customer advising. In order to improve safety, citizens called for more supervisory staff and revitalization of train stations.

The ÜSTRA Citizens' Report, which with its 12 planning cells reached a comparatively large number of participants, also makes the method's weak points clear: its implementation is time-consuming and therefore cost-intensive as well. The citizen reporters, who overall drew mainly positive conclusions, criticized the amount of labour and time pressure, and the lack of direct influence on the implementation of proposals.[82] Overall, however, the method proved its value: the resulting Citizens' Report comprised several hundred practical recommendations for improved LPT in Hanover.

4.5.5.2 Citizens' Forums in Baden-Württemberg The Academy for Technological Impact Assessment in Baden-Württemberg received an assignment in 2001 from the Verband Region Stuttgart (Association of the Stuttgart Region) to create an applicable participatory practice to allow citizens to evaluate the factors of functionality, interactive design, service-friendliness, and user-friendliness of the transport services offered (intermodal services, parking space management, PT connection information systems, a virtual town hall). 'Anticipating possible conflicts and problems in accepting the developed and still-developing technological measures, one of the [Verband Region Stuttgart's] essential intentions was to work with future users to create solutions and to adjust the development plans accordingly.' [83]

80 Stiftung Mitarbeit (Foundation for Cooperation) (1996), p. 45.
81 See also Böhme et al. (1998).
82 See Böhme et al. (1998).
83 Academy for Technological Impact Assessment in Baden-Württemberg (2002), p. 47.

The Citizens' Forums were carried out following the planning cell method (see Section 4.5.4). The events, however, did not take place within four consecutive days; instead they took place during four evenings (6:00 pm to 10:00 pm) at intervals of one per week. Participants judged this timeframe to be convenient. The results of the citizens' forum were reviewed by members of the municipal council and the city administration and were in part implemented.

4.5.5.3 Citizens' Exhibition From August 2005 to June 2007 the project 'mobikult' was sponsored in the context of the European LEADER+ initiative and carried out by the Centre for Technology and Society at the University of Technology in Berlin in cooperation with the nexus Institute for Cooperation Management and Interdisciplinary Research GmbH, as well as partners in six Brandenburg regions. The starting point of this project was the concept that a mobility structure reflects the attitude of all stakeholders involved in the mobility process. It addressed the issue of how mobility service providers organize this structure, how politicians discuss it, how planners conceive it, and how customers use it. In order to manage, conduct and influence these attitudes, a high level of communication and moderation processes is necessary.[84] In the context of the LEADER+ cooperative project 'mobikult – mobility is culture' several intersectoral forms of cooperation for public mobility were created.

The project provided an impression of how innovative cooperation strategies can shape public mobility offerings in a customer-oriented and efficient way even in economically underdeveloped regions. In order to implement such a project, it is necessary to consider the specific situation on-site. Alongside local circumstances such as the size of the city or already existing transport services, this includes basic social conditions. In the 'mobikult' project, the Citizens' Exhibition practice was used to communicate citizens' thoughts and ideas.[85]

4.5.5.4 The Future Search Conference: participation process for local transport in Düsseldorf On the occasion of the Düsseldorf Future Conference, there was a controversial discussion about a planned new underground train line through the downtown area. Rheinbahn, the transport company active there, reacted by inviting the citizens to engage in a dialogue. In 2002, the opening session took place. A citizens' forum was instituted where several sessions were held to prepare a final Future Search Conference. In preparation for this conference, there was an effort made to describe Düsseldorf public transport in terms of eight main interests.

During the course of the two and a half days of the Future Search Conference, 53 people worked mainly in small groups that met in varying

84 Schmidtmann (2006).

85 See www.mobikult.de/echo.html (accessed September 2007).

arrangements. The participants agreed to a common vision for the future and finally developed recommendations for actions that could achieve this ideal.[86]

The moderators of the Düsseldorf Future Search Conference refrained from providing information input through expert lectures. The focus was placed on participants' individual perspectives, which were visualized during the work process.

4.5.5.5 The Transport Forum in the City of Heidelberg The Heidelberg Transport Forum was initiated by the mayor for the conception of a city-wide transport development plan in 1991. Over a period of two years, a research group composed of representatives of various organizations (interest groups as well as political parties and citizens' initiatives) met one to two times per month. In addition, public gatherings of citizens took place. The goals were to develop a local information exchange, and to create recommendations for action in improving the transport situation.

Approximately 70 people took part in each of the meetings of the research group. Members of the administration and external experts made their specialized knowledge available and sought to incorporate the research group's proposals into their work. 'The two-year work process included, among other things, a comprehensive appraisal of the transport problems in Heidelberg, the presentation and evaluation of prospective solutions, the creation of a transport mission statement, and various packets of measures for the tasks of the transport forum.'[87] The results of the work were revised by a private planning office. In 1994, the municipal council passed the transport development plan.

4.6 Interest Representation Institutions – Who Speaks for the Passengers?[88]

4.6.1 User Associations – Common Features and Experiences
Holger Jansen, Martin Schiefelbusch

The term 'passenger [or 'user'] association' in the following text means *independent* groups that are, initiated *by users* of public transport services, and *supported* organizations that represent the interests of these groups. Alongside passengers' concerns, they address other issues to some extent – this will be discussed in more detail later. In contrast to the other organizations presented in the following sections, passenger associations are independent, unlike an advisory council (Section 4.6.3), which is organizationally attached or subordinate to a transport company. On the other hand, this independence

86 See Wachholz (2001).

87 Wachholz (2001), p. 15.

88 See authors' names in the section heading below.

means that they are dependent on their own means to carry out their activities and cannot,[89] like the British watchdogs (Section 4.6.5), fall back on long-term sources of financing. Volunteer commitment and the motivation if members are thus important resources.

4.6.1.1 Public transport in the context of the cycles and emergence of automotive transport Transport politics after World War II and the economic boom years of the 1950s and 1960s were, in many countries, heavily oriented toward individual transport (Table 1.1). The constant construction of road infrastructure and the yearly record numbers in newly licensed vehicles define the discussion in transport politics. The mission statement was to create a car-friendly city. In the Federal Republic of Germany, the Leber Plan,[90] introduced in 1967, contributed to the goal that no one should be more than 20 km away from the next motorway.

Individual transport has been consistently increasing. The market share of LPT has thus decreased dramatically, even though passenger revenue has to some extent remained constant – with differences between urban and rural regions. In response to this, many transport companies have withdrawn, particularly in terms of supply. In rural regions, the only thing remaining has often been a minimum service for students and customers who do not have personal vehicles of their own.

Even with all of the pleasure that many people experienced in travelling with automobiles, the negative side of constant transport growth has not gone unnoticed, particularly since the 1970s. The environmental movement gained more and more supporters in the second half of the 1970s. Along with many other issues, this movement addressed transport. No one could continue to ignore environmental damage through exhaust, road construction in previously undisturbed landscapes, noise, soil sealing, the high number of victims in road traffic, and many other dark sides of constant transport growth.

How were the players in the political/administrative system to deal with these developments? Some sought to contain the negative sides of transport with technological means, including the promoters of emission purification. Others advocated a speed limit not only on city streets but also on highways.

Where car transport problems were particularly prevalent – in urban centres, but also along major long-distance transport connections – measures were increasingly taken to protect residents and the environment.[91] An active roll-back of MIT – unthinkable a few years before – was seriously discussed at the end of the 1980s, the more so as some cities and regions had had positive

89 At least in most cases, one-time or special-purpose support from outside is certainly possible.

90 Named after the former Federal Minister of Transport Georg Leber.

91 For this topic see Reiter (1999); Schmucki (2001); Lundsgaard-Hansen (2000).

experiences in this area. Cities that set an example include Zurich, Bologna, and Groningen with their transport freezes; Oslo and Bergen with their early congestion charge and their later car-free spas and tourist regions.

The discussion constantly returned to the question of what alternatives to individual transport could be offered. Interest in public transport thus increased significantly in comparison to previous decades, and also lead to investment in modernization and construction. There were still recognizable differences in the strategies and priorities applied in transport politics. In revitalising PT, Zurich and Switzerland in general once again served as an example; Freiburg/Breisgau, Karlsruhe and Strasbourg, however, played a role as well.

On a broader front, however, limitations in MIT were not able to take hold. Even though development of public transport was able to contribute to maintaining and increasing passenger revenue – a definite success in light of concurrent growth in motorization – it was not able to drastically change the modal split.

Since the mid-1980s, the transport turnaround[92] has significantly lost popularity as a goal for politics and planning. More difficult basic economic conditions have contributed to transport being propagated as a precondition for economic development. MIT limitations have thus been applied in an essentially conservative manner. Instead, transport companies have made integration and cooperation their goal. With in the broad acceptance of this goal, there exist significantly differing opinions.

Public transport is not (any longer) fundamentally called into question in this process. In the course of European integration and economic megatrends, existing PT structures, however, are being scrutinized. The model followed almost everywhere in Europe of PT as a social service provided through public companies in insular markets (see Section 1.1) has started to receive pressure to justify itself. Especially since Great Britain began its far-ranging deregulation and privatization of its transport markets in the 1980s, the EU Commission has been propagating competition for the market, especially as it has occurred in Scandinavia and France.[93] As part of this process, transport contracts have started to be formulated with more exact service descriptions and contract auditing. Together with general efforts to economize public funding, this has created high efficiency and effectiveness requirements for PT. This has made for a more difficult position, for representatives of passenger interests as well.

4.6.1.2 Passenger representation through the decades The history of passenger associations in Germany is short compared to associations in many other

92 Hesse (1993).

93 After long discussions, a compromise was finally achieved with Ordinance 1370/07, which continues to allow independent production of transport services through public companies, if it fulfils specific basic requirements.

countries. Some environmental associations or organizations in the field of sports or leisure have been in existence for more than 100 years. General consumer organizations can likewise look back on many decades of active work. In contrast, specialized organizations that represent the interests of customers of public transport first started to develop at the beginning of the 1980s. The longer tradition of such organizations in the United Kingdom (Section 5.7.4) is an exception compared to other countries.

The emergence of passenger associations is thus closely connected with the situation of public transport described in the previous section. From the start, two courses of action have intermingled:

1. *transport political* concerns deal with strengthening PT in competition with automobiles;
2. *consumer politics* deal with representing users' interests in relation to service providers.

It is characteristic of PT that both goals overlap in the work of associations, which is caused by the structure of the industry. Public transport in its early stages was shaped by companies with national monopolies. In rail transport, the market was dominated by the national rail businesses, which often played a strong role in regional bus transport as well through subsidiary companies. The existing transport operations in larger cities were generally public property.

Companies generally reacted to declining passenger counts with further service cutbacks. A fatal vicious cycle emerged: passenger counts continued to decline because of service cutbacks. In many cases, the end was shutting down routes. A proactive approach oriented toward winning customers was sorely missing, because neither the companies nor their owners had any interest in it. Service providers themselves were more interested in going so far as to weaken offerings that seemed uninteresting to them through unattractive timetables – at least according to a frequent allegation by transport associations.

The automobile lobby at this time had been organized for decades. As an example, the Allgemeine Deutsche Automobilclub (German Public Automobile Club (ADAC)), which was founded in 1903, was purportedly named as a representative of numerous other organizations. There were no interest representation groups for customers of public transport. What could the passengers even hope to achieve? On one side was a group of committed passengers, on the other side were the powerful national monopolies who could afford to be indifferent to passengers' opinions.

Nevertheless, more associations and organizations gradually formed with the goal of actively representing passengers' interests. There are three recognizable phases for this development in Germany, Austria, and Switzerland.

Late 1970s and 1980s – first attempts: the transport politics of the 1980s were shaped by an ideologically charged debate.[94] Discussions about forest decline or catalytic converters shaped the discussion about the future of automobiles. In light of the negative effects of automobile transport, the neglect of other transport providers was increasingly criticized. Deficiencies in local situations or intentions to make further cutbacks in PT often gave those affected the impetus to get involved. Up until today regional initiatives are a main pillar of passenger interest representation (see Section 4.6.1.3); a few organizations that act on the national level have arisen alongside them. One of these, the Verkehrsclub Deutschland (Transport Club of Germany (VCD)), was able to achieve one of the first major successes on the national level with its campaign for a discount card for train tickets. After several years of campaigning, in 1992 the federal railways introduced such a discount card for all customers, with which anyone could travel at half price. Previously this discount was only available to youth and seniors.

The relationship between passenger representatives, companies, and politics was particularly difficult in the early years, even if the history of passenger associations may seem to have been shaped by fewer tough conflicts than other citizen movements that arose at the same time. Gradually some companies began to rethink their positions, and they became receptive to their customers: The first documented passenger advisory council in Germany was founded by the Kölner Verkehrsbetriebe (Cologne Transport Services), in 1989.[95] The first discussions about reforms for the future of the federal railways had begun.

The 1990s – expansion of the scope of work: the 1990s were shaped by steady growth in associations. Increase of membership numbers, construction of infrastructure in the former GDR, and new campaigns and ideas were on the agenda. Important milestones for the associations included the rail reform of 1994 and the regionalization of local rail transport in 1996. The federal government transferred local transport to the states and provided substantial funding for this process. The federal government thus attested to the fact that local rail transport is an essential public service for rural regions and cannot be provided without subsidies.

Numerous new transport authorities for LPT on the national level recognized the associations as valuable partners and incorporated them into their work. Exchanges in passenger advisory councils (or boards) and in informal dialogue became increasingly extensive. Passenger associations experienced significant advantages through these new responsibilities, particularly in regional rail transport. The federal railway was no longer in charge as a major administration, instead the country Minister of Transport, or a lower office, was responsible in each respective state. Between the associations, a mostly constructive and businesslike working atmosphere

94 See Klenke (2007) and previous section.

95 See Fliegel and Schiefelbusch (2005) and Section 4.6.3.

evolved in many places. Most states founded transport or fare federations. This produced new opportunities for passenger associations to participate.

From the year 2000 to the present – consolidation: new umbrella organizations and new issues – this is how developments starting in the year 2000 could be described. Deutsche Bahn's plans for a new price and fare system comprised a large part of the work. Similar to air transport, passengers were to receive a discount if they made an early commitment to a particular train connection. Deutsche Bahn had to change central elements of their price system due to public pressure, in which the associations played no small part. Further issues included passengers' rights as customers in cases of delay or cancellation (see Section 2.6). Associations compiled arguments and presented them, for instance in hearings at the German Bundestag. In recent years, the planned capital privatization of Deutsche Bahn has shaped the discussion. The picture on the local and regional levels is dominated by preservation and development of offerings under increasingly difficult basic economic conditions – that is cutbacks in subsidies. European transport and consumer politics has also gained more attention. Platforms for a continuing dialogue among associations have developed on the national as well as the European level, where the European Passengers' Federation was founded in 2002 (see Section 4.6.2). Membership numbers of most organizations in Germany have been stable or slightly declining during this period. There have been other issues – particularly the employment market and the effects of globalization – that have moved people.

4.6.1.3 Who speaks for the passengers? – passenger association archetypes The spectrum of organizations that represent the interests of PT users has developed out of the particular situations of individual countries, and thus has taken differing forms. In an international comparison, it can however be seen that there are certain models that appear consistently. Not only 'passengers-only' associations work to represent passengers' views. There is a wide range of other organizations that are also active in this field:

Passengers-only associations are groups that exclusively focus on PT and its users. They feel committed to the *overarching goal* of preserving and improving public transport and thus do not engage in prevention- or deflection politics that arise from local concerns alone. In their work there is often an implicit connection with improving their own position as PT users, as well as with the transport political goal of winning new customers.

Local Groups (citizen initiatives, committees) play an important role because regional, concrete issues such as the preservation of a specific route or improvements to timetables often provide the momentum for action. The advantage of such groups thus lies in the relative facility of mobilising stakeholders due to the immediacy of their concerns. However, they have serious disadvantages as well. These groups tend to lack a view of the public transport system as a whole. In many cases, they dissolve after their goal has been achieved. The opportunity of approaching certain goals through long-term work is thus often not used.

For *environmentally-oriented associations*, the design of an environmentally friendly transport system is the main priority. Their goal is to shift transport from automotive transport forms.[96] From this perspective, PT is supported because of its ecological advantages; there is, however, similar attention paid to bicycle and pedestrian transport. Just as conflicts in planning can arise between the interests of bicyclists and those of PT, these organizations can also be forced to set their priorities for support.

This also applies to *nature conservation associations* in a more specific sense. While they do recognize the fundamental advantages of PT, they also criticize PT-related building activities just as they do road-building projects if they encroach upon flora and fauna. They do, however, have interesting potential as passenger allies: The issue of nature conservation reaches other major groups and thus gives the associations access to substantial resources, and a privileged place in planning procedures as officially recognized interest representatives.

In a similar manner to how nature conservation associations understand themselves as advocates of nature, the goal of *group organizations* is to represent specific selections of the population. They do this irrespective of any particular situation, and they act as passenger representatives if their group is part of the PT customer base. The primary focus, however, is always the interests of the group as opposed to the needs of travellers at large, which can easily lead to conflicts with other organizations. In terms of the customer structure of PT, the most common forms of these institutions include student/parent organizations, associations of seniors and disabled persons, and to some extent workers' unions as well (as representatives of commuters).

As a counterpart to these organizations with narrow fields of focus, general *consumer associations* have the goal of representing consumers on a very broad scale. The role as passenger is thus one among many; they must accordingly be competent in numerous fields in order to successfully represent their interests. This can cause smaller organizations to come up against their limitations. In addition, conflicts of goals cannot be ruled out if they want to represent users of automobiles as consumers as well, even if users of personal motor transport have their own organizations in most countries.

In a narrow sense, *transport historical organizations* do not represent passengers' interests. They are more focused on the attractiveness of the train[97] as an artefact, often with a concentration on historical aspects. They regard the development of a modern transport that is responsive to contemporary

96 In recent years, shifting from aeroplanes to ground-based PT has become a significant further goal.

97 Including street trams. Similar interest groups for buses are possible, but are significantly more rare in practice.

needs with a mixture of indifference and scepticism. However, it should be remembered that there certainly can be overlaps in the people involved in such groups and those in passenger associations. Train museum organizations are of interest insofar as they are able to mobilize volunteer action on a significant scale for the provision of transport services. Outside of this field such empowerment is significantly more rare.[98]

This characterization shows the differing perspectives. In reality, the various forms mix in terms of how the individual organizations view themselves, their memberships, and also the courses of their development. Furthermore, there are temporary as well as long-term forms of cooperation that span many organizations.

A decisive element for the success of associations on the transport political stage is the relationship of these forms to one another, and what kind of impression this makes in terms of their presentation. In the best case, a trustful, eye-to-eye dialogue can take place in which the associations' work can at least partially compensate for the failures of the market mechanism. Contrary developments can arise for many reasons; alongside reluctance from service providers, there can also be doubts that the association actually represents the views of the travelling public.

4.6.1.4 Methods and functioning of associations According to Hirschman (1970), consumers' interests can be primarily expressed through 'exit' (the choice of another provider in cases of dissatisfaction) and 'voice' (the articulation of requests and (dis)satisfaction). The 'exit' option is usually not available as long as there is no other means of transport (automobile, bicycle) available. In the area of 'voice', there are several identifiable *tools* that possess a varying degree of formalization and institutionalization:

1. communication between individuals or groups and the administration, policy-makers, companies, media;
2. informal, internal discussions;
3. political lobbying;
4. demonstrations, campaigns, etc.;
5. public relations;
6. participation in round tables, Agenda 21 groups, etc.;
7. case-specific, informal citizen participatory practices (that is, Future Workshops);
8. participation in parliamentary hearings and boards (that is, as an 'expert citizen');
9. petitions and referenda;

98 The main example is the citizen's buses established in the rural areas of The Netherlands and Germany. There are also some examples in recreational transport (Schiefelbusch (1995)).

10. citizen participation in the context of formal planning practices (as long as it is provided, that is, for infrastructural projects).

In principle, passenger associations can make use of all of these methods, as long as they have the necessary resources (especially time for active members) and formal access to the procedures. It should be emphasized that the significant industry-specific planning and authorization procedures in many countries – including Germany – only provide partially guaranteed possibilities for exercising influence.[99] The 'voice' option, too, is thus not directly available, but must first be developed with the construction of suitable forms of dialogue and politically effective presentations of the need for action. The prospects for success thus depend on the receptiveness of the opponent. Also, this constellation leads to particularly high demands on the work of consumer and/or passenger organizations themselves.

What methods should passenger associations apply for their work? First, the respective target groups should be differentiated. For passenger associations, this essentially includes their own members, political and administrative decision-makers, employees of transport companies and federations, the media, and members of partner or umbrella organizations. One of the ways that political persuasiveness is created is when associations can gain credibility that they speak for a significant portion of the population and/or represent goals that have widespread popularity. The *search for allies* is thus an important part of political work. The representatives of passenger interests can find allies and support in various other organizations, the most important of which are presented in Table 4.2.

Table 4.2 Other interest representatives and potential allies

Supporters in environmental groups	Representatives of individual groups of passengers	Other organizations
Pedestrian organizations Cyclist associations Associations for the advancement of environmentally friendly transport in general Nature conservation organizations	Local citizens' initiatives/ organizations (as representatives of the passengers of one town, etc.) Workers' unions/work councils (as representatives of commuters) Student representatives and/or parent organizations Seniors' organizations Disabled persons' associations	Project-specific citizen initiatives Commercial enterprises that profit from PT Social services/welfare associations

Source: authors' depiction.

99 Schiefelbusch and Bodensteiner (2005).

Transport companies, too, can be passengers' allies in issues such as gaining political acceptance for new services or acquiring public funding. If there is good informal contact between service providers and passenger representatives, the latter can even be used as an 'intermediary' if the company does not want to put forward the initiatives themselves due to political considerations.

Overall, however, it is also true here that common interests can only be reached to a certain degree, particularly because the basic constellation of 'provider–user' can be expected to lead to polarities. Further difficulties are presented by the tight connections between politicians and companies. In this situation, companies tend to give particular consideration to political demands and sensitivities that can go so far as adopting positions that place their own business at a disadvantage.[100] Passengers who do not want to accept such decisions thus end up advocating the company's own interests, even if contrary to the company's officially stated position.

User organizations must always be aware that every group pursues its own goals and sets its own individual priorities. The common advocacy of one thing that all groups see as important is thus a wise means in the political process. However, a group will get support from another group only if it does not present disadvantages for their own constituency. This must be negotiated on a case-by-case basis.

Probably all associations publish member newsletters to *disseminate information* among members. These publications serve to inform their members. Some associations offer their publications in bookstores or send them to larger distributors.

Although transport political discourse often leans towards focusing on the concrete, rationally describable advantages and disadvantages, the emotional components of mobility should also be considered in the argumentation (cf. Section 1.2.1).

Usually only larger associations with full-time employees organize big campaigns. There are several issues for which only larger associations with full-time employees can organize campaigns. A strictly volunteer association can generally not organize comprehensive campaigns due to lacking human and financial resources. Through publicly effective actions, associations try to position their respective subject matter together with current transport political issues. The most important disseminators are members of the media, who are able to give subject matter a wide reach through the mass media. Likewise, professional exchange with other associations or active individuals in politics and the administration has a high value.

100 One example would be when routes and stations are chosen that are not justifiable from a transport perspective only to avoid being contrary to politicians' personal preferences.

4.6.1.5 Conclusion and prospects for the future The history of passenger associations in Germany is comparatively short. Only in the 1980s did various organizations begin to form. Membership numbers stayed comparatively low. This certainly may also be due to the fact that passengers and consumers in general do not have what is necessary for an effective organization: a common and therefore organizable interest.

For strictly environmental associations there is a different picture. The preservation of natural conditions is more clearly communicable than the common interest of passengers. The imbalance of personnel and financial resources between companies and interest representatives should also not be overlooked.

Associations found success particularly in the 1990s. A few years after the start-up period, many active groups were accepted and included in political and company processes. During this time, many passenger advisory councils formed, whose numbers would grow to the hundreds in German-speaking countries (see Section 4.6.3). The following points should be named as essential deficiencies:

- passenger and consumer associations have no legal claim to participation. They are dependant on the good will of service providers;
- the imbalance of important financial resources between associations and service providers is still unchanged. Only larger associations have enduring and ensured funding through membership fees and contributions. At most, public funds are available for particular occasional projects;
- a high value must be accorded to an association's independence and to the stabilization of what content their work addresses. For example, for the Schlichtungsstelle Mobilität (Mobility Arbitration Board) or the Deutschen Regionaleisenbahn (German Regional Railways) their work for passengers is tightly connected to their close cooperation with transport and mobility service providers, which can encroach upon this independence.

Overall, the work of the various associations has brought positive results. They have succeeded in pointing out alternatives to individual modes of transport over a long period of time. Many different topics have been addressed and different target groups could be reached. Some activists have also become professionals – this may be a loss to the respective association, but can be a way of changing the providers' attitude substantially. Another positive factor is that all of the associations have lasted and continue to endure through decades of changing basic conditions. It has been possible to continually win new members for work that is mostly shaped by volunteer commitment.

The impact of the associations' work is difficult to quantify. They have succeeded in some cases and failed in others. But like in any political process, it would be misleading to consider just the outcome of a particular debate. Decisions are usually the result of recent and past discussions and experiences, and it is not always possible to establish precisely who was the first in putting forward a certain argument. On the whole, transport user associations' work has certainly contributed in many places to keep and raise interest in public transport and protect it from further decline.

4.6.2 The European Passengers' Federation: Integrating the Users' Views across Countries
Trevor Garrod, Holger Jansen

The European Passengers' Federation is a young organization. It was founded in October 2002 as a non-profit association under Belgian law. It is a federation mainly of voluntary bodies but also includes two statutory organizations. Currently it has 26 member bodies in 15 countries and contacts in several others. Some associations are purely for rail passengers (for example, Deutscher Bahnkundenverband, Railway Development Society/Railfuture, Association luxembourgeoise des amis du chemin de fer). Most are concerned with buses and trams and sometimes ferries as well as trains (for example, Bond van Trein-Tram- en Busgebruikers, Federation Nationale des Associations d'Usagers des Transports, Associazione Utenti del Trasporto Pubblico).

4.6.2.1 Starting points Several factors led a number of existing national or regional bodies to start working together in the 1990s. Firstly, there was concern over cross-border traffic. It was becoming easier and easier to cross national borders by road, especially after the Schengen Agreement. However, crossing by rail did not get easier and sometimes indeed became more difficult. Associations in Alsace, Baden-Wuerttemberg, Luxembourg, Switzerland and Austria started working together and organized a series of conferences in such places as Luxembourg, Mulhouse and Lauterbourg.

No one woke up one morning and declared 'Today I shall form the Eureopean Passengers' Federation'. Instead, an evolutionary process took place as in various countries users realized they needed to talk with each other, work with each other and then have a more structured manner of working. Following informal contacts in 1998 and 1999, and a meeting in Zurich, two lunches were organized in the French city of Lille in July and November 2000. These were attended by British, Belgian, French, Dutch, Luxembourg, German and Austrian colleagues who agreed to call a Conference of European Rail Passengers in Paris in April 2001. That event attracted some 30 members and was addressed by Dr Ellwanger of the Union International des Chemins de fer. It considered in particular high-speed rail and how it could be best integrated with conventional rail. Everyone wanted to make this an annual event and so they met again in Brussels in May 2002, when nearly 50 people attended. In the meantime, the European Commission organized a hearing in Brussels in October 2001 to enable passengers and consumers to voice their opinions on international passengers' rights and obligations. Many user representatives attended this event which led to further contacts with the European

Commission's Directorate for Transport and Energy. By now it was also clear that a more formal structure was needed. Discussion papers were circulated to 13 associations and bodies who came together in Ghent in October 2002 and agreed a set of statutes to set up the European Passengers' Federation.

4.6.2.2 What are EPF's strengths? Most of the associations have existed for at least 20 years and campaigned in their own countries or regions for better public transport. Indeed, the two statutory bodies – the Rail Passengers' Council and the London Transport Users' Committee – go back much further. So they have a lot of collective experience upon which to base their work.

Secondly, independence is a strength. The Federation's money comes from members, individual and corporate. It is not funded by political parties or trade unions. If the member organizations receive cash from outside bodies such as local authorities or the railway industry, it is for specific projects, such as a conference, publication or piece of research. This also applies to the European Passengers' Federation itself. EPF welcomes sponsorship for specific projects or purposes; but currently it is funded entirely by its members and from profitable events such as conferences which it organizes.

Thirdly, the members are public transport users, whether or a daily basis commuting to work or on a less frequent basis for business and/or leisure. Some associations organize surveys to gather data on the good and bas aspects of the services which members use. Sometimes they undertake research into the costs and benefits of a new or enhanced service. In larger countries especially, such as the United Kingdom, France and Germany, our national associations have a network of local branches and affiliated local users' associations who provide them with detailed local knowledge.

Lastly, EPF is a democratic body. Its policies are determined through informed discussion from below, and not imposed from above. Members also now use the Internet to exchange opinions and help formulate policies. In brief, EPF members aim to be intelligent, well-informed, realistic laypersons.

In 2003 EPF issued a 13-point manifesto in English, French, German, Dutch, Italian and Esperanto. The 13 points sum up the results of the early discussions and it was not difficult to agree them – since all of the associations had, to a greater or lesser extent, had similar experiences.

4.6.2.3 Current tasks for EPF *International long-distance trains* are an important part of EPF's work. A working party makes efforts to promote long-distance services. It has started an on-line bulletin 'Initiative Fernverkehr', in German and French. An important aim is to promote integrated ticketing – 'one fare, one ticket for one European journey'. At times where many stations can no longer sell international tickets, this is a particular challenge. EPF representatives have addressed this in meetings with companies like Eurostar and Thalys.

Responsible rail: by this we mean a rail system that is safe, looks after its customers, compensates them for delays and cancellations or has contingency plans when things go wrong. EPF representatives have had useful dialogue about this matter with bodies like the European Infrastructure Managers and Community of European Railways, as well as the European Commission and Members of the European Parliament.

A level playing field: in recent years, much has been spoken and written about the unfair advantage which airlines have over railways. Many of these go back to arrangements such as the Chicago Agreement of 1944, which was made to encourage what was then a young industry. Is this still necessary? That is the sort of issue which EPF raises with politicians and the European Commission. The issue will not be easy to tackle, but it must not be ignored. It is part of a wider scenario of having a level playing field for rail to compete with other modes. Compensation for late trains is important as well as a requirement for operators to publish punctuality figures. But there should also be comparable requirements for bus, coach and ferry operators. EPF takes every opportunity to raise such issues with national and European politicians.

To sum up: the European Passengers' Federation is a young organization. But it is eager to work with political decision makers, transport professionals, academics and others with whom it can share common ground. It already has much experience in our individual countries. The work at European level is a natural extension of that experience.

4.6.3 Passenger Advisory Boards – Concepts and Experiences in Germany and Austria

Martin Schiefelbusch[101]

4.6.3.1 Introduction Passenger advisory boards (Fahrgastbeiräte (PABs)[102]) are the most frequently used form of 'formal' transport user involvement in the German-speaking countries. The first such board was set up in Cologne in 1989.[103] Today, about 100 boards exist,[104] most of them set up since 1995.

Characteristics of passenger advisory boards include:

101 Based on Fliegel and Schiefelbusch (2005).

102 Other terms used are *Fahrgastforum*, *Kundenbeirat* and *Nahverkehrsbeirat*, but these are also used for other forms of user involvement. In this chapter we refer only to 'user/passenger advisory board' as the most common expression. As an English term, 'passenger advisory board' is mainly used.

103 Bröer (1994).

104 Figure based on an Internet search in summer 2004. No such boards were identified in Switzerland. The precise number of boards is difficult to establish because groups which do not meet the above-mentioned definition occasionally use the same term as well.

- institutions with an *advisory role*;
- consisting wholly or mainly of *public transport users* and/or *members of organizations* representing user interests;
- set up on a *long-term basis* or permanently;
- to *discuss and express* the *users' views* on public transport, both on current matters and future plans;
- they have some formal '*foundation*' due to their links to an operator or transport authority.

The latter point distinguishes them from the activities of user associations (Section 4.6.1) or civic action committees who have no such formal backing and guarantees of being heard.

Despite these common features, their specific form varies considerably, both in terms of their formal and organizational framework (for example institutional affiliation, statutes) and their scope and approach. This section gives an overview of the different concepts currently in use.[105]

4.6.3.2 Experts' or lay users' board? Two main types of PABs can be distinguished by their members' backgrounds: So-called lay users' boards on the one hand and experts' boards on the other. The latter are also referred to as association advisory boards, which already points to the main characteristic of these organizations: they are composed of representatives of organized groups, for example, passenger associations, organizations representing senior citizens or young people – usually each organization has one member on the board. Conversely, lay persons' boards consist of people without an institutional affiliation who are chosen from personal applications – sometimes a link to an association is even a criterion for exclusion. Mixed boards can be found as well.

The choice of one of these models implies different foci of their work, based on their members' backgrounds:

- Representatives of organized groups argue more from a 'political' perspective and aim to direct attention to their organization's interests. They often have a strong personal interest and built-up knowledge through other activities in their association and the information these provide to their members (through magazines, events etc.). Compared to lay users, they have higher expectations of their work on the board. A common aim is, for example, to discuss at eye level with the operator's representatives.

105 If not stated otherwise, information in the following section is based on the survey described in Fliegel and Schiefelbusch (2005), where 37 out of 98 contacted organizations participated.

- On the contrary, lay users' boards tend to argue based on their members' personal experiences. Local or even individual views are more prominent than a general or strategic perspective. This applies both to their own concerns and to the feedback they give to topics raised by the providers.

A crucial challenge for both types of organizations is to identify particular interests – be they personal or group-specific ones – and prevent them from dominating the debate. Professionals tend to view 'experts' boards' more critically in this respect, due to their ambitions. However, some prefer the professional exchange offered by this concept.

Lay users' views are often appreciated as 'genuine customers' views'. Over time, however, even lay users are prone to a certain narrowing of their interests and/or the development of an 'expert's attitude'. Therefore in some boards, membership is limited to a number of years or terms.

The survey identified only eleven passenger advisory boards where no members had affiliations to transport-related pressure groups. This is much lower than in an earlier survey in the late 1990s.[106] However, it is not possible to establish whether the associations' members were selected for the board because or despite of this fact. Both appear possible.

4.6.3.3 Structures and activities Most boards have between 20 and 30 members. A balanced representation of different socio-demographic groups on the board is essential to achieve a good overview of the users' views. Actual public transport use is most often quoted as a necessary criterion. Other groups named are people with disabilities and younger people (Figure 4.8). However, due to the structure of applications received, some groups may in the end be over- or underrepresented on the board.

Passenger advisory boards are mainly set up on *district level* (35 per cent) or for the *operator's catchment area* (27 per cent). Sixteen per cent are set up for one city only. Transport authority districts and German countries account for 8 per cent each.

The costs of running the passenger advisory board are typically met from the marketing or public relations budget of the operator or authority with which it is affiliated. Of those who responded to this question, 43 per cent quoted costs of less than €500 per year (Figure 4.9). An expenditure of more than €5,000 was given only twice. Higher costs arise when a board is moderated externally. In some cases, the working time spent by the operator's staff on the board meetings may also have been considered as cost to the company.

The boards deal with wide range of *topics*. The meetings are dominated by discussions on network and timetable development which is on the agenda of 95 per cent of the respondents (Figure 4.10). Infrastructure (stations), quality

106 Meyer-Liesenfeld (1997, 1997a).

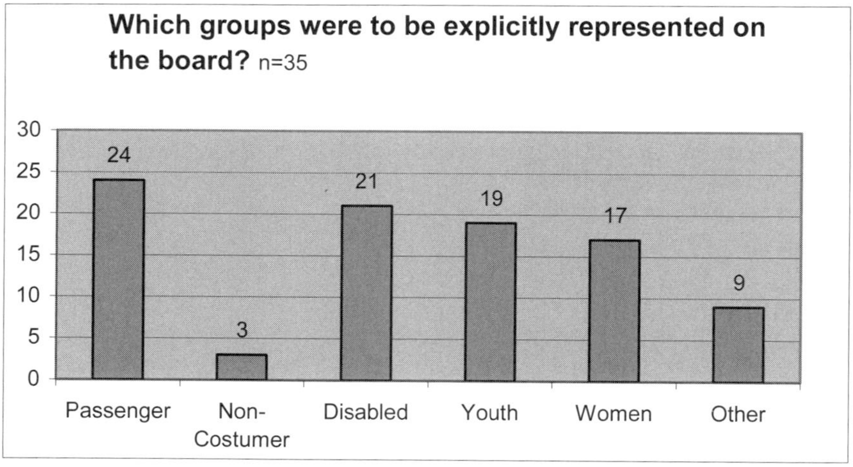

Figure 4.8 Criteria for the selection of board members

Source: Fliegel and Schiefelbusch (2005).

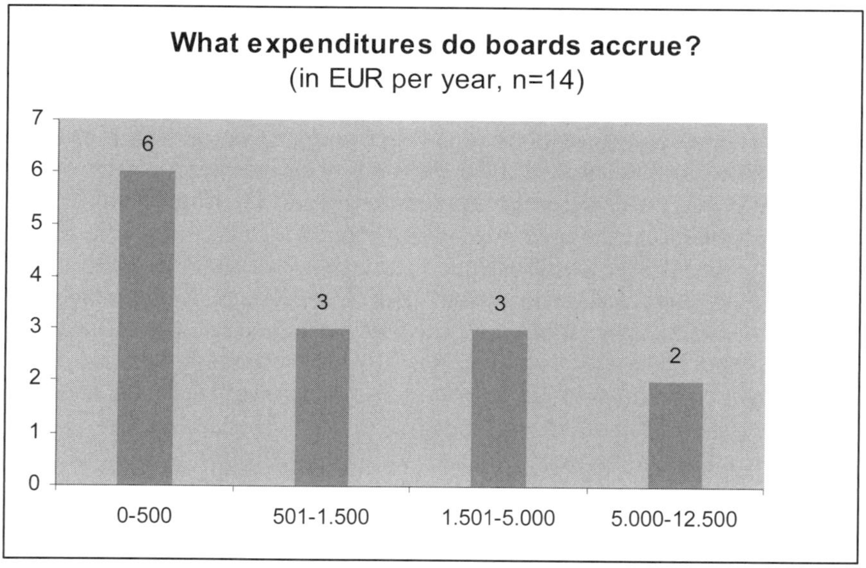

Figure 4.9 What expenditures do boards accrue?

Source: Fliegel and Schiefelbusch (2005).

standards and fares are named by three quarters of those questions. Vehicle design, carriage of bicycles and other types of building work were named by half of the boards. Representatives of the operator/authority are usually present, but their role is limited to presenting proposals and answering questions. Occasionally on-site visits and other activities are also part of the board's work.

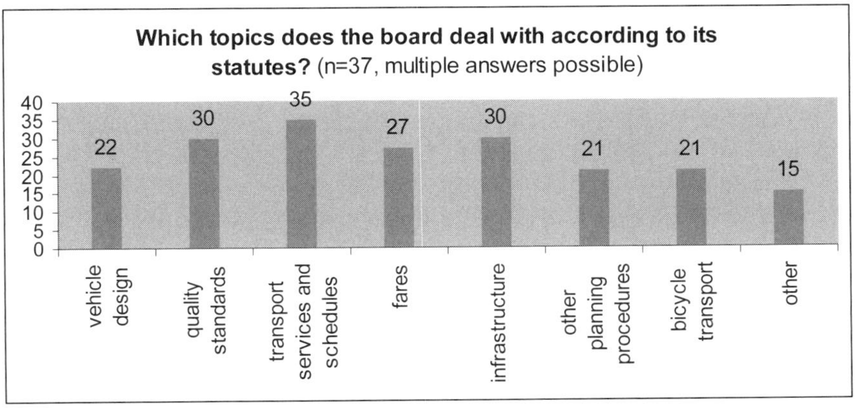

Figure 4.10 Issues addressed by user advisory boards

Source: Fliegel and Schiefelbusch (2005).

Another important aspect is the position given to *individual complaints* on the boards' agendas. This refers to cases of service deficiency or other individual problems experienced either by the board members themselves or reported to them.

On average – and probably based on a quite rough estimate of the respondents – half of the suggestions made by the board were *accepted* by the operator/authority. However, individual values vary greatly, and lack of success is evidently more common in those areas which account for the majority of the boards' work: fares, infrastructure design, network planning and timetables, and capacity enhancements.

An involvement of passenger advisory boards in *other institutions or processes* is another way of exercising influence. However, the boards surveyed rarely participate in formal planning procedures concerning public transport. Only 17 per cent said they did so, giving discussions on public transport master plans and infrastructure (station) projects as examples.

Documentation and *external relations* use similar means to recruit members. One third of the organizations use one or more of the options given in the questionnaire (minutes of meetings, press releases, newspaper reports, websites and annual reports). But it is also worth noting that one third of them still do not publish their work at all, and less than half of the organizations hold their meetings completely or in part in public.

4.6.3.4 Results Looking back, the evolution of passenger advisory boards now allows an evaluation of their effects on public transport planning and operator policy. The respondents' impression in 15 cases was that collaboration with the operator had developed in a positive way. However, the majority of the respondents (19) had not noted any change in either direction, and three perceived relations as worse than they were when the board was set up.

Future expectations voiced include above all continuation and expansion (12 nominations each) of the boards' activities. Seven wanted a greater engagement of their members and a better user involvement in general. Four saw publicity activities as an area due for improvement.

In the future, recruitment of new members may become harder. Problems to attract volunteers were already mentioned by several respondents, usually with reference to younger age groups. These difficulties can also be observed in other areas that depend on civic engagement.

To overcome this problem, it is necessary to understand (potential) members' *expectations* and cater for them in the board's work. Experience from other areas indicates that this is less a matter of financial or material interests. Rather, it is important to recognize and value the participants' idealistic motivations – in this case their willingness to contribute to better public transport. This means that operators and authorities as sponsors of the board have to involve the board in a meaningful way – in particular to consult in good time and take its suggestions seriously, on both small and big issues. User advisory boards are often funded from the provider's marketing budget, but it is not sufficient if the companies only see them as a publicity channel.[107]

Recognition can also come from the outside world, for example, from fellow passengers. To do so, the board's work must be visible to outsiders. Hence expansion of their publicity activities, which are to date often limited, should be considered, but agreed upon by the members involved in order not to demand too much of them. Operators have mixed views about this, because discussions within the board can become less open and uninhibited as a result.

A key characteristic of a working board is a trusting relationship with the other stakeholders involved where different views can be handled in a constructive way. Such a relationship often only develops over time, but then it is useful for the providers as well.

107 Rahs (1996), p. 448.

4.6.4 *OV-Platforms – Working Together for Public Transport*
Huibert Verdoold[108]

4.6.4.1 Formal background In The Netherlands, administrative responsibility for regional public transport was transferred to the regional governments (12 provinces and seven city regions) in 2001 with the Passenger Transport Act 2000 (Wet personenvervoer 2000). The aim of the transition of powers has been to create a more costumer-focused and cost-effective market in regional public transport. The 19 transport authorities are responsible for public tendering procedures and award operation franchises, among other tasks.

To involve passengers in service planning, the Passenger Transport Act gives consumer organizations a leading role. They are supposed to represent the interest of customers of within a franchise area. The regional transport authority and the franchisee must ask the consumer organization(s) for formal advice on crucial aspects of the franchise or the operations. The law does not stipulate which consumer organization should be talked with. It only gives three broad requirements for assigned consumer organizations, leaving the decision who to talk to basically to the authority and operator. These requirements are that organizations must be a legal entity, they must attend to the interests of public transport users and, lastly, organizations must be active on a national or regional scale. Assigned consumer organizations are interlocutors for both regional transport authorities and transport companies.

The *authority* has to ask advice of the assigned consumer organizations about the requirements placed upon companies through tender descriptions and obligations associated with a franchise (PVE in Dutch). To give these organizations a good opportunity to offer advice, the authority has to organize a consultation with the organizations and must ask for their advice in good time. After the advice has been given, the consumer organizations must be informed of the decision. As well as asking advice about the PVE the authority is obliged to inform the consumer organizations about changing policy, new infrastructure and maintenance work.

Now the section looks at the formal relation between the assigned organizations and the one or more *transport companies* who take care of the franchise of an authority. The Passenger Transport Act names a number of areas on which transport operators have to ask advice the organizations' advice, namely:

- execution of the timetable;
- information to travellers about timetables and fares;
- transport conditions;
- tickets;
- availability of tickets (when and where to buy your ticket);

108 Based on Verdoold (2005).

- options of payment;
- accessibility;
- social safety.

If laid down in franchise conditions, they must also ask about:

- how to deal with complaints;
- compensation for delay;
- quality targets.

Just like the authority, a company has to hold a minimum number of meetings with consumer organizations. Also, formal advice requests must be made in good time. Furthermore, within six weeks a company has to give a written and reasoned reaction on the advice received. If they wish not to follow it, consumer organizations may claim a second consultation. After that, the company is allowed to implement its decision.

4.6.4.2 Organization There are about 25 OV-platforms, which is used as a generic term for these organizations which have different names from region to region. Although the Act has three requirements for assigning consumer organizations, in practice all 19 regional transport authorities have made their own choices about which consumer organizations to talk to. In early all regions, there are three main areas of representation:

1. ROVER (organization of public transport users);
2. organizations for the handicapped and chronically ill; and
3. organizations representing the elderly people.

In some regions environmental groups, cyclists, labour unions and groups representing youngsters and/or students also participate. Sometimes individual persons – not acting on behalf of an organization – also take part in the discussion. This example makes clear that the number of organizations and persons with which authorities and companies have to discuss varies. In practice, it is also a question of supply and demand.

In course of time, most authorities have chosen to institutionalize the discussion because they wanted to deal with a formal consultative body or just with fewer parties. Two main forms are distinguished. First, consultative bodies that are formally part of the authority and function as advisory boards. In the second form, consumer organizations have organized themselves in an association or a foundation and the authority has decided to assign that organization as interlocutor. If we look at the resources of the OV-platforms we see large differences. One platform receives €250,000 a year and so have their own office and employ two persons. This is an exception, however. Most platforms receive an annual budget of around €10,000 or even have no budget at all. In this case they can often declare expenses and are reimbursed on a case-by-case basis.

4.6.4.3 Experiences As the previous section has shown, there are many different ways in which regional institutionalized user participation is presented. However, the most important questions are: Does it work? Will the voice of the users be heard? Do OV-platforms obtain anything with their advice?

In general, it is a growth process for all parties concerned, consumer organizations as well as authorities and companies. Here are some points:

1. In the first place, working together in OV-platforms is sometimes hard for the volunteers that represent the different organizations. They have to be team players and not see other members as a threat to their own interests.
2. A member of an OV-platform represents an organization. While forming an opinion, this person needs to be fed by the parent organization. But in practice the relation between local member and parent organization is often rather weak.
3. In a more general sense one may ask: are the interests of *all* users heard in OV-platforms? It is a problem of representation. Most travellers on public transport are teenagers going to school and students. However, many OV-platforms have no proper representation for this group. It appears difficult to find young people who want to represent their fellows. There are different reasons. Most young people do not have the opportunity because most meetings are held during the day. In addition, most students study in a region for a relatively short period. However, the major problems are the structure and culture of customer participation. It is not seen as 'cool'.
4. There is a big competence gap. For example, members of ROVER know about timetables and route planning, whereas a member of a cyclists' group is generally more interested in safe roads for cyclists or cycle parking locations near tram or bus stops. In real life this sometimes leads to inefficient meetings and frustration. As a solution some platforms have set up sub-committees to divide the work. The sub-committees prepare advice and after that the whole OV-platform will discuss it and draw up their final advice.
5. As mentioned earlier, some OV-platforms are so poorly supported by their authority that no-one expects them to do well.
6. Authorities and companies do not always realize that members of OV-platforms are mostly volunteers and not professionals, so that it is important to submit comprehensive papers and ask for advice in a timely manner. Most authorities and operators have struggled with the latter.

7. Public tendering of transport franchises is also a new activity for regional authorities, operators and consumers alike. So, for instance, consumer organizations had to learn the hard way when the best time was to influence decision-making. Often, their formal moment of advice was granted after all major decisions had already been taken.
8. If an OV-platform functions well and the authorities are happy with the quality they receive, they are inclined to give it more tasks and send it more papers. So overflow of activities threatens the OV-platforms.

4.6.4.4 Conclusion: working together in OV-platforms: does it work? Regional user participation is on the right track, although there are still OV-platforms which are lagging behind. Their trip has just begun. Some 'roadblocks' have already been mentioned. However, there are some positive examples that show of the success of OV-platforms:

1. most of their advice is followed up;
2. both transport operators and authorities take them seriously;
3. the members are enthusiastic and skilled;
4. they have enough facilities to do their work;
5. they are working together in a common goal: improving public transport for all.

The successful OV-platforms benefit users of regional public transport. It is the task of OPC as national support organization to support, stimulate and help weaker OV-platforms. The strong ones must maintain their high level of quality.

4.6.5 London TravelWatch

John Cartledge[109]

4.6.5.1 Status and tasks In the United Kingdom, official consumer representative bodies – so-called watchdogs – have been established by law since the late 1940s to protect consumer interests in areas where normal market rules do not apply. Such bodies exist for all of the major utilities or regulated industries: post, telecommunications, gas, electricity, water, airlines, airports, railways and (some) buses. In London and its commuter hinterland, the equivalent body in the passenger transport sector is now known as London TravelWatch. It has four principal spheres of activity: consultation, monitoring, investigation, and representation.

Consultation involves it in a continuous dialogue with the bodies which plan and operate the services. They are formally required to seek its views before making any major changes to routes, times or fares. Railway lines and stations

109 See also Cartledge (2005).

cannot be permanently closed without the relevant regulator's agreement, and this cannot be given until he or she has received and considered a report from London TravelWatch, taking account of any representations made to it by those affected. But in practice, consultation extends over the entire range of issues which affect the users.

So London TravelWatch is also routinely involved in discussions about, for example, ticketing systems, passenger information, vehicle design, station facilities, safety (that is from accidents), security (that is, from crime) and the particular transport needs of young travellers, or women, or people with disabilities. It does not wait passively to be asked for its views, but asserts them proactively in regular meetings with the decision-makers in the industry. And it has a research programme leading to widely-circulated reports on issues of current concern: recent topics have ranged from the views of disabled rail users to the principles of good complaint-handling practice, and from the intelligibility of timetables to the quality of toilets at transport interchanges.

Monitoring means that London TravelWatch keeps a close watch on the quality of service provided. This includes both 'hard' measures of reliability and 'soft' measures of user satisfaction, for example with cleanliness, information and staff helpfulness. With 24 train companies, 10 Underground lines, two light rail systems and more than 700 bus routes to scrutinize, this is no small task. Where trends are favourable, the watchdog is always ready to give recognition to those who deserve it. But when services deteriorate, London TravelWatch is vigilant in seeking explanations and remedial action.

Investigation is required when users are dissatisfied with any aspect of the service received (or, quite possibly, not received) and when in the complainant's view the operator's initial response has been inadequate. In these circumstances, London TravelWatch is the appellate body to which the aggrieved user can turn. It handles some thousands of such cases each year and if it believes that a complaint has merit, it will seek a fuller explanation, a more fulsome apology, a more convincing assurance of future improvements and (where appropriate) more generous compensation. But it is not only concerned to obtain more helpful responses from operators to complainants. London TravelWatch also requires to be satisfied that the root cause of the problem which gave rise to the appeal is being seriously tackled.

Representation is provided by London TravelWatch in the wider arena of public debate. Because it is the recognized voice of public transport users, its views on any issues which may affect them are sought as a matter of course by national and local government and other public agencies. So it is regularly commenting on proposed changes to laws and regulations affecting transport and consumers, on land use planning and traffic management policies, and on matters of safety (for example, it is always represented at public inquiries into serious accidents). It gives evidence to parliamentary investigations, and is represented on a variety of official advisory bodies. As a leading member of the European Passengers Federation, it is becoming increasingly involved with the European Union's agenda, as this extends into the spheres of transport policy and consumer protection.

4.6.5.2 Structure and resources London TravelWatch has 13 members of its governing Board, who are recruited by public advertisement and serve as volunteers. They are expected to devote two or three days per month to their role and they receive payment of £4,000 (approximately €6,000) per year. The Board meets in full session every month, but its members are also required to attend a number of sub-committees, working parties, public consultation events, site inspections and the like. The Chair of London TravelWatch spends three or four days a week on its work, and receives commensurately greater remuneration.

Members of the Board are appointed by the London Assembly, the elected body which scrutinizes the work of the Mayor and approves his budget. The Assembly also funds London TravelWatch. Members normally serve for an initial two-year term, but are reappointed if they make an effective contribution to its work. When making appointments, the Assembly must follow guidelines to guarantee the probity of its decisions and ensure that there is a proper balance by age, gender, ethnicity, physical ability or disability, and place of work and/or residence. All members must be regular users of London's passenger transport systems.

The Board is assisted by a professional secretariat which now numbers 22 staff (full-time equivalent). They are divided roughly equally between caseworkers, policy and research officers, and administrative support workers. As in any public body, ultimate power of decision-making rests with the Board members, but much of the detailed day-to-day work is delegated to the professional officers who provide them with advice and implement their decisions in accordance with the policy guidelines they have determined. Figure 4.11 shows the structure of the organization and the distribution of tasks.

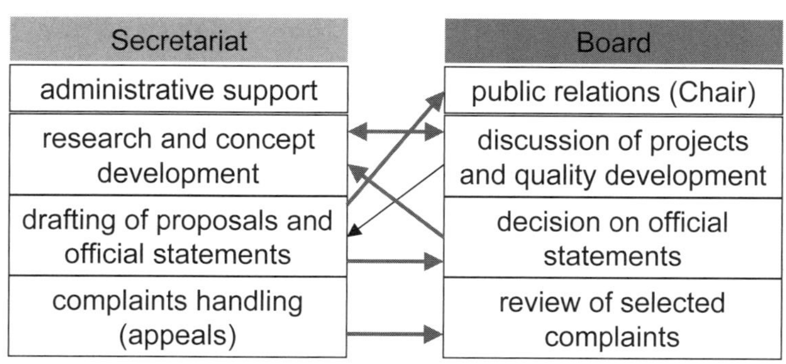

Figure 4.11 Organization of London TravelWatch

Source: based on Staudinger and Schiefelbusch (2004).

4.6.5.3 Impact There is no scientific proof of the effectiveness of a pressure group. It is always possible (however improbable) that every improvement in London's transport for which London TravelWatch has campaigned over the years would have occurred without its existence. But it is noteworthy that when London TravelWatch routinely polls all of the aggrieved passengers who seek its help, two out of three give its efforts a positive verdict – no small achievement when it is remembered that as an appellate body it deals only with the most litigious complainants and/or the most complex cases.

London TravelWatch is not an executive body, and can only operate by dialogue and persuasion. Therefore, its policies must be well-researched and its arguments must be convincing. The operators are not obliged to accede to its wishes, but they are obliged to respond to the questions it raises. And because its work is conducted in public, the world at large can judge which side of any argument has the greater merit. London TravelWatch's meetings are open to all, its reports and papers are freely available on its website (www.londontravelwatch.org.uk), and its activities are widely covered in the media.

4.6.5.4 Open questions The British system of institutionalized consumer representation has gone through several changes of government and by and large been maintained and expanded. When parliament enacted legislation in 1999 to create a new system of city-wide government for London, it not only reaffirmed London TravelWatch's role but extended it to cover additional modes of transport including cycling, walking, taxis and ferries. However, one should not forget that London TravelWatch also faces some difficulties in its everyday work.

The first problem is that when dealing with complaints from users, London TravelWatch has *no formal power*. It cannot impose its wishes on operators and it does not even have the right to demand the disclosure of documents or the attendance of witnesses. The transport providers do have to answer its questions, but how they do so is for them to decide, and if they wish to be uncooperative or uncommunicative, that is their choice. There are particular difficulties in handling complaints about the behaviour of individual members of staff, because employment law may restrict the information that can be publicly revealed. Because London TravelWatch cannot instruct the operators, it has to persuade them – and this means that it must always be careful to express itself in diplomatic terms.

The second problem is one of *political accountability*. London's regional government, the Greater London Authority, consists of a Mayor and an Assembly, both of which are directly elected. The Mayor has executive power and controls the transport authority. The Assembly's role is primarily to hold him or her to account, and to scrutinize the policies of the various agencies which he or she appoints. So the Assembly continually holds hearings about, and issues reports on, aspects of transport policy and delivery. But it is also – now – the body which appoints and funds London TravelWatch. Because there is no clearly drawn line between its agenda and its responsibilities on the one hand, and those of the watchdog on the other, the opportunity for rivalry and tension is obvious.

The third problem relates to the organization's *modal remit*. Originally, it was concerned only with championing the needs of public transport users. But when London's administration was reorganized, the government changed the remit of London TravelWatch – against its wishes – to include all the users of transport systems (other than freight) for which the transport authority is responsible. This means that its constituency now includes not only cyclists and pedestrians – whose needs are broadly complementary to those of passengers – but also motorists, of whom this is not necessarily the case. How to resolve any competing priorities which may arise from this situation is a question which can be challenging both for members and officials.

The fourth and last problem can perhaps best be summarized as one of *democratic credibility*. When London TravelWatch was first established (under another name) more than 50 years ago, it was perfectly acceptable for such an advisory body to rely entirely on the experience and judgement of its own members and advisers, and its legitimacy was not challenged. It did not have to engage in 'outreach' activity. But today it exists in a different world, where advances in information technology have created new opportunities for instantaneous communication and popular participation.

It is possible for millions of citizens to vote directly, using mobile phones, in a ballot to decide the winner of the Eurovision Song Contest. Can such developments be exploited to capture public opinion on transport policy issues? What value should be attached to the results of polls conducted without serious prior debate? What are the strengths and weaknesses of referenda? Or any of the myriad of other techniques of public consultation?

And how does it deal with the problem of asymmetry in the debate? By this is meant the fact that the voices of the few who believe that their interests and privileges are threatened by a particular proposal (such as the introduction of a bus lane or a change in the stopping pattern of a train) are invariably heard more loudly and assertively than those of the many who it is intended to benefit but who have much less incentive to speak up on their own behalf. These are questions with no easy answers.

4.7 Conditions for Success and Limitations for Participation

Christian Bodensteiner[110]

As previously described in Chapter 4, there are several models and methods for participatory practices. An important question for the proper implementation of a participation method is the *character of the project*. The problems that

110 I would like to thank Martin Schiefelbusch for additions.

are solved through project planning and participatory practices can be divided into two categories: 'problems of innovation' and 'problems of conflict'.[111]

The main tasks of participative approaches to problems of *innovation* are to search for and invent new, user-friendly solutions. Thereby, participants' creativity should be activated because a normal citizen's knowledge, experiences, and the ideas are frequently not used in internal administrative planning.

With innovative project design, practices can inspire the creativity of the largest possible number of participants. Since new ideas are particularly in demand, it is necessary to enable as many participants as possible to use the full force of their creativity. Informal participatory practices can be used for this purpose, as described in Section 4.5.

Figure 4.12 ranks various forms of participation in terms of how new and how complex their topics are. The newer the participatory practice is, the more creativity is demanded. In this respect, as higher demands are placed upon participants, they must be more intensively supported.

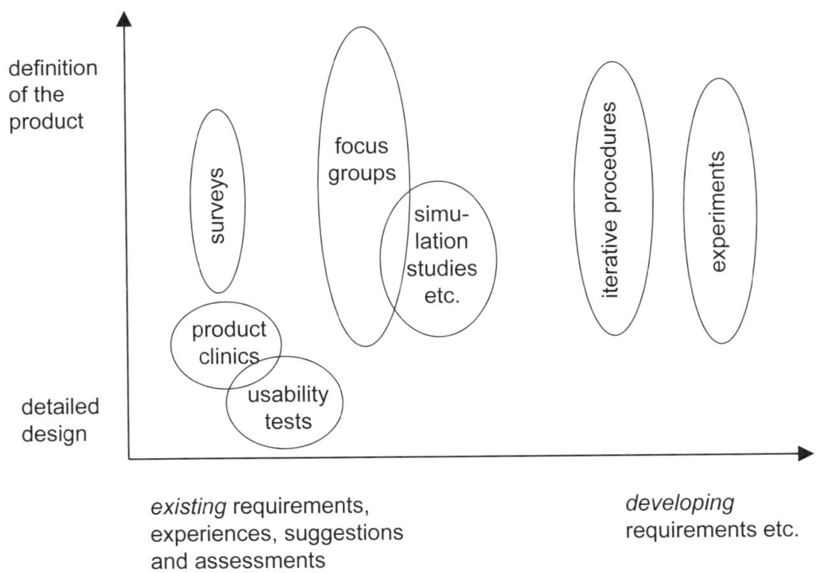

Figure 4.12 Suitability of user participatory practices for different topics

Source: Konrad and Truffer (2003), p. 9.

The main task of participative planning for planning problems involving *conflict* is the development of collectively binding decisions. Planning often involves the allocation of scarce resources and thus creates 'winners'

111 Dienel (2005).

and 'losers'. In such cases, participation can be used to reach a sustainable compromise or at least to make the general parameters for such decisions more transparent and comprehensible.

Beyond formal and informal participatory practices, the political realm offers further possibilities and necessities for participation – if the former are missing, having a say in the political process takes on special significance. Section 4.2 summarized the instruments available for such work. However, the summary also makes clear that an effective representation of interests in this realm requires skills that are not easily available to everyone. Project- or topic-oriented participation, like the approaches described in Sections 4.4 and 4.5, create a 'free space' that opens up a wider sphere for collaboration.

Citizen and user participation can, as seen in this chapter, have an effect on how planning is carried out in public transport as well. With the implementation of participative methods, for example, problems can be identified which would not otherwise be recognized. Furthermore, the right kind of participation can also inspire participants' creativity, which leads to new and unexpected ideas, or at least lead to an increased understanding of planning decisions. Finally, if varying perspectives enter into the discussion, citizen and user participation can give both voice and importance to the opinions of those concerned.[112]

However, participation in public transport can also peter out if the general parameters do not make sense or if they are insufficiently applied from the beginning. It is still striking how few cases there are of mandatory public participation in the design of transport networks and timetables. Certain groups are involved especially in local transport planning, partially due to the support of LPT laws, and partly due to voluntary decisions on the part of transport authorities. In contrast, the design of local transport services is primarily in hands of transport companies – in the numerous steps and practices involved in the planning, there is no space provided for cooperation with or listening to the public.

Participation offers advantages and possibilities, but there are of course limits set on citizen participatory practices. They are thus 'neither a cure-all nor arbitrarily applicable management tools.'[113] The success of the actions taken in planning transport depends to a significant degree on public acceptance.

An investigation by the German Federal Ministry for Transport, Building, and Housing[114] examined the conditions for success for LPT in rural areas.

The main findings concerning the implementation of new mobility concepts demonstrate that success is not possible without key figures: 'The solutions must be based on clear, local desires and initiatives and social supervision. In addition, they require first and foremost a dynamic *personality as a key figure.* Otherwise they do not function, there is no interest in them and they soon

112 Cf. Konrad and Truffer (2003), p. 7–8.
113 Bischoff et al. (1995), p. 148.
114 BMVBW (1999).

disappear.'[115] This experience can also be applied to citizen participation: the results of participatory practices are just plans at first. Without something more, their implementation is not certain, especially for informal practices. They require identification, or at least recognition of the results from the people who will carry them over into everyday life. For this reason, alongside the identification and empowerment of stakeholders and key local figures, *official support* is of great importance to the success of participatory practices.

In conclusion, the examples from local transport planning show that participatory practices can be successfully established and carried out. The future is open for them to be put to more use.

115 BMVBW (1999), p. 52.

Chapter 5
Planning Public Transport, Customer Representation and Passenger Rights in Europe

5.1 Austria

Holger Jansen

5.1.1 Political and Geographical Context

Austria is a federal republic. The country comprises 83,871 km^2 with a total of 8.28 million residents (2006).[1] The capital is Vienna (Wien) with approximately 1.65 million residents. The federal level of the Parliament consists of two chambers: the National Council (the lower house) and the Federal Council (the upper house), which represents the states. The head of the Federal Government is the Federal Chancellor, who is also the Chairman of the Federal Cabinet, which consists of 12 ministers. The National Council is elected every four years.

There are parliaments in each of the nine federal states (countries). Elections for the country parliaments take place every five to six years. The party with the most votes appoints the governor. The administrative unit below the country level is called a district. Districts are lead by a district commissioner who is appointed by the country government. The district is administered by its district commission. There are 15 statutory city districts and 84 political districts, making 99 in total. The smallest administrative unit are the 2,358 (2006) municipalities. Austria has been a member of the European Union (EU) since January 1995.

5.1.2 The Transport Market and the Organization of PT

The Bundesministerium für Verkehr, Innovation und Technologie (Federal Ministry for Transport, Innovation and Technology (BMVIT)) is responsible for transport. In the country governments, departments for transport coordinate the political duties surrounding means of transport. In Vienna, a branch within the Magistracy is responsible for transport. The states allot part of the operative

1 Statistik Austria, http://www.statistik.at, accessed August 2007.

duties to transport networks and the Verkehrverbundorganisationsgesellschaft (transport authority organizational association (VVOG)).

Non-profit services in local and regional transport in particular are federally funded.[2] The states take responsibility for and coordinate public transport. The municipalities organize purely short-distance local transport. The legal basis for cooperation between federal, regional and municipal transport networks and transport companies is presented in the law regarding the Organization of Local and Regional Public Transportation (Ordnung des Öffentlichen Personennah- und Regionalverkehrs (ÖPNRV-G 1999)). The law came into effect on 1 January 2000.

Accordingly, the federal government is required to 'assure the provision of a core offering for public, regional, and local rail transport in the scope of the services ordered or delivered in the timetable year 1999/2000'.[3] For existing transport contracts for buses (effective from 1 June 1999), the federal government assumes operational costs not covered by fare income.[4]

5.1.2.1 Transport Associations and transport network organizational associations At present there are eight transport associations in Austria:

- Verkehrsverbund Ostregion (VOR), Vienna;
- Verkehrsverbund Niederösterreich/Burgenland, Vienna;
- Steirische Verkehrsverbund, Graz;
- Oberösterreichischer Verkehrsverbund (OÖVV), Linz;
- Salzburger Verkehrsverbund, Salzburg;
- Verkehrsverbund Tyrol, Innsbruck;
- Verkehrsverbund Vorarlberg, Feldkirch;
- Kärntner-Linien – Verkehrsverbund Kärnten, Klagenfurt.

The BMVIT[5] describes the legal construction of a transport network as follows:

> A transport network is a cooperative institution operating through contracts made under private law between regional authorities (primary and financing contracts between federal authority and the respective state) as well as between the regional authority and the individual transport companies (service level agreement). The financing regional authorities (federal, respective states and,

2 The Federal Government made c. €870 million available for this purpose in 2004. Deficiency in receipts through special fares for commuters or trainees are excluded from this figure.

3 Section 7 ÖPNRV-G.

4 Section 10 ÖPNRV-G.

5 http://www.bmvit.gv.at/verkehr/eisenbahn/nahverkehr/verkehrsverbuende.html, accessed September 2007.

> as the case may be, respective municipalities) are represented in a management committee. The management committee's duties are regulated in according bylaws. Among the most important duties are the formulation of general, political parameters for the transport networks as well as clarification of the financing of the network's activities. The transport companies taking part in the respective network are organized in a cooperative alliance of transport companies wherein they are responsible for carrying out network activities.

Transport networks are meant to orient themselves to the conditions of each state. The VOR includes parts of Burgenland and Lower Austria. The transport networks determine a common fare, unified dispatch systems and ensure transport quality. Transport networks are the platform for transport companies to coordinate with one another. There is a transport network organizational association (TNOA) in every state. The BMVIT[6] describes this as follows:

> The transport network organizational association serves as a mediation authority between the regional authorities that finance the network and the transport companies involved. In this respect it is the central clearinghouse and carries out inter-company planning. The network organizational associations are purely regional entities in Austria.

The duties of a TNOA are named in Section 18 of the ÖPNRV-G:

- specifications for the fares within the network;
- coordination of transport services;
- arbitration board and clearing house for revenue sharing;
- recommendations to regional authorities for local and regional transport planning;[7]
- announcement of services, processing of transport contracts.

5.1.2.2 Transport companies The largest transport company in rail transport are the Federal Austrian Railways (Österreichische Bundesbahnen, ÖBB). Long-distance and local transport are organized by the ÖBB Personenverkehrs AG (ÖBB Passenger Transport plc) which is a subsidiary of ÖBB Holding. The company started carrying out functional business on 1 January 2005. The legal basis for the rail reform was the Bundesbahnstrukturgesetz (Federal Railway Structure Law) from 2003. The sole owner of the ÖBB is the federal

6 http://www.bmvit.gv.at/verkehr/eisenbahn/nahverkehr/verkehrsverbuende.html, accessed September 2007.

7 The legal basis for planning is Section 11 ÖPNRV-G. States and municipalities should cooperate with transport networks in the planning and altering of demand-oriented transport services.

government. The ÖBB operates a total of 4,000 trains per day, of which c. 3,600 trains are local transport. 22 further rail companies are active in the country. They are all publicly owned.[8] In the year 2005, the ÖBB had c. 191.6 million passengers.[9]

The 22 other rail companies served c. 22 million passengers. The largest bus company is 'Postbus', which belongs to the ÖBB. The company possesses over 2,100 buses, operates 860 lines, and has 3,970 employees. In 2005, 244.8 million passengers used their bus service. A total of 605 million passengers use public bus transport. The largest municipal transport company is Wiener Linien. They employ 7,750 workers and serve c. 772 million passengers (2006).[10] The market share for public transport (modal split) is 35 per cent in Vienna (nationwide average: 17 per cent).

5.1.3 Transport Planning and Participation

The General Transport Plan exists on the federal level. It was introduced on January 25th 2002. The transport infrastructure is meant to be constructed with an investment program of €45.1 billion. The goal of this is to strengthen Austria as a business location. The existing transport networks are thus intended to be expanded according to their needs. Rails receive €29.9 billion of the funding for this project. The BMVIT created the plan together with country governments, the Österreichische Bundesbahnen (ÖBB), federal infrastructure businesses and legal interest groups.[11] There was no direct citizen participation.

Every country created their own plan for the transport development. In Vienna, the 'Transport Masterplan' has been in effect since 2003. Participation opportunities for the citizens have been expanded in the past. In 1970 and 1980, the planners included transport experts. In 1994, there were consultancies with representatives from the public. For the 'Transport Masterplan' decided upon in 2003, there was a comprehensive participatory practice. It was oriented toward all citizens. The practice began in May 2002 in the individual districts. Following intensive discussions, there were closing presentations in early 2003 and a closing event in June 2003.

Mediators supported the discussion process between citizens and experts. The planners organized c. 200 sessions in the districts, with external

8 http://www.parlinkom.gv.at/pd/steno/PG/DE/XXII/NRSITZ/NRSITZ_00056/ SEITE_0192. html?P_PM=SEITE_0192, accessed September 2007.

9 Annual report of the ÖBB.

10 http://www.wienerlinien.at/WienerStadtWerke/DOWNLOAD/Betriebsangaben_2006.pdf, accessed September 2007.

11 More information at http://www.bmvit.gv.at/verkehr/gesamtverkehr/generalverkehrsplanung/gvp_alt.html, accessed September 2007.

support.[12] At the beginning, there were preliminary presentations. Following that were 'open space moderations' in work groups to determine topics with two to four meetings per work group. The experts of the Transport Masterplan participated in the closing presentation in the districts. The current status was accessible on the Internet as well. The following was to be found there:[13]

> The treatment of material takes place in work groups whose activities are determined in three conferences. The technical community and the districts are invited to collaborate in advisory boards which will become informed about the progress of the project and can have an influence on the formation of the Masterplan through statements about its content. Technical and scientific assessment and commentary are to be assured through an international expert workshop. Finally, the results of the citizens' participation will be incorporated into the mobility team in the work process as well.

For the entire practice, the Magistrate used c. €300,000. Of the citizens' recommendations, the planners incorporated 25 percent in the Master Plan. 20 percent were dropped as a 'conflict of goals', a further 55 per cent were outside of the authority of the individual districts of Wiener Linien. The Österreichische Bundesbahnen (ÖBB) and Wiener Linien, as the largest transport companies, are proposed to develop the remaining planning in Vienna and the surrounding areas.

5.1.3.1 Passenger Advisory Board Wiener Linien Wiener Linien has had a passenger advisory board since 15 March 2004. On that day, the constituent session of the new board took place with 16 members. There were 600 applications for the board. The members of the advisory board work on a volunteer basis. The chairman and speaker of the passenger advisory board is Professor Hermann Knoflacher. The advisory board is composed of various socio-demographic groups:

- students (and/or parents, 2);
- 19–25 years old and employed (2);
- 19–25-year-old students (2);
- 26–60 years old and employed (2);
- 26–60-year-old unemployed homemaker (2);
- 60+ retiree/pensioner (2);
- people with reduced mobility (2);
- non-users (2).

---

12 The districts were grouped into so-called district tandems. They consisted of one, two or three districts.

13 http://www.mobilinwien.at/mpv/verfahren, accessed September 2007.

Organized interest groups are not represented in the advisory board. The basis of the board is a designation in the transport contract between the city of Vienna and Wiener Linien. The following was printed in a press communication[14] from Wiener Linien in regard to the first session of the passenger advisory board on 15 March 2004:

> The passenger advisory committee expressed a desire to not only get information about Wiener Linien's concrete and planned projects, but at the same time to get information about all other plans that are relevant to public transport users.

The passenger advisory board meets four to five times per year. The members advise in regard to, among other things, an improved offering. Among the determined adjudications is the fact that Wiener Linien is required to report results. The sessions are not public. The chairman represents the board for external inquiries.

Among its activities to date, the board has addressed the topics of cleanliness, improved information about everyday travel and about disturbances. Thus, at the incitement of the passengers, vehicles are cleaned during the day as well. In addition to local topics, the City of Vienna informs the passenger advisory board about new transport plans in the city. Information about timetables is to be improved and to be present upon entering the train platform.[15]

Wiener Linien conducts customer surveys and offers a customer hotline. By their own account, the feedback from these inquiries has a high value. Wiener Linien surveys their passengers regularly for which they work together with the institute Socialdata. The results are described as follows:[16]

> In c. 40 inquiries, people's position toward public transport is communicated. Alongside the categories of offering, reliability, provision and maintenance of information, conduct of the personnel, safety/personal feeling of safety and comfort, aspects about the image of public transport and price-performance were also addressed. In the category 'safety/personal feeling of

14 http://www.ots.at/meldung.php?schluessel=OTS_20040316_OTS0184&typ=, accessed September 2007.

15 Further information about the topics that the passenger advisory board has addressed can be found in the magazine *24 Stunden für Wien* by Vienna Public Services at progs.wiennet.at/24online/n175/html/175_19_L.html, accessed September 2007. The year named as the committee's founding year, 2003, is however not correct.

16 The study compares several cities in Europe and is called BEST 2001. More information can be found at at http://www.wien.gv.at/vtx/vtx-rk-xlink?DATUM =20020124&SEITE=020020124010, accessed September 2007. For informations about the Passenger Advisory Board see http://www.wien.gv.at/vtx/vtx-rk-xlink?SEITE=020 031128015&DATUM=20031128&S0=socialdata#P0, accessed September 2007.

safety' Wiener Linien holds the leading position, as last year, and now four times in a row. Wiener Linien was also the best in its group in the category of 'comfort'. The second rank was reached in the areas of 'transport offering', 'information', 'value/price-perfomance relationship' and 'customer care'. Ranking third were the areas 'punctuality/reliability', 'conduct of personnel' and 'overall satisfaction'. The category 'social image' was in the fourth rank.

5.1.3.2 Passenger Advisory Board ÖBB

The basis of the passenger advisory board is the ÖBB's customer charter for long-distance transport. A dialogue with customers is provided in this document. The ÖBB introduced a customer forum which met for its constituent session on 28 September 2006. The board is lead by Professor Hermann Knoflacher. Twenty people belong to the advisory board as volunteers who were chosen from c. 2,800 applications. Representatives of diverse customer groups are members of the forum. Selection criteria were gender, age, occupation, state, and relationship to using the train. They thus represent Austrian train customers. The ÖBB understands the customer forum as a link between passengers and the ÖBB. The members express desires, needs, concerns, and criticisms. The board meets about four times per year. Topics of the sessions to date have included traveller information, treatment of customer inquiries, mobile phone reception in trains, cleanliness, support of passengers in wheelchairs, and improvements to the infrastructure.

5.1.3.3 Transport concepts and participation possibilities in the states In Lower Austria, the Niederösterreichische Verkehrskonzept determines the details of transport development. It was created in 1997. The Verkehrsverbund Nieder-österreich/Burgenland or the Niederösterreichische Verkehrsorganisationsgesellschaft (NÖVOG) handle inquiries or concerns from individual customers. The Chamber of Labour (see below) in Lower Austria is not satisfied with participation possibilities in Lower Austria. They are seen as narrow and/or poor. The transport companies present no form of participation for passenger or customer organizations.

Burgenland, like Lower Austria, is closely connected with the Vienna country in terms of transport. The transport organization covers part of Burgerland and Lower Austria. The remaining portions are served by the Vienna-based VOR. The presently effective transport master plan ('Gesamtverkehrskonzept Burgenland') was released in May 2002.[17] The mobility needs of the public

17 In the state government, the Administrative Department of Land Use and Regulation, the Department of Total Transport Coordination is responsible for this. The full concept is accessible at http://www.aufguteverbindung.at/images/content/verkehr/gvk_burgenland.pdf, accessed September 2007.

and the economy are meant to be at the centre of transport politics. The country government intends to do this with the construction of transport infrastructure.

In *Styria* there is a transport master plan of the Country of Steiermark ('Gesamtverkehrsprogramm des Landes Steiermark')[18]. Transport organizations and TNOAs work with the user group Fahrgast Graz/Steiermark, particularly on electronic timetable information. The position given to user and passenger organizations depends on the respective occasion. Many contacts exist on an informal level. Organizations are presently not institutionally included in discussions. The transport network leads a yearly market research study in which a customer barometer is determined. The results serve primarily to check the marketing measures within the transport network.

In *Upper Austria*, a transport master plan has been in effect since 2005. Since 1998, a separate department has coordinated the planning of public transport.[19] The state's strategy endeavours, on the one hand, to prevent transport companies from further thinning out the offering. On the other hand, public transport in the region is to be addressed, organized according to needs and co-financed through the region, that is through appropriate concepts and regional mobility management.

The country government intends to make public transport more attractive through construction of railway lines, a city S-Bahn, and electrification. The trams in Linz are to be expanded. According to law, TNOAs and transport networks must be integrated in the planning process. Other stakeholders are included according to their degree of involvement. There is no institutionalized participation or legal claim for organizations. The Oberösterreichische Verkehrsverbundgesellschaft (OÖVG) determines details with the respective stakeholders.

Since 1999, there has been a user advisory board instituted by the country in LPT for Linz. Various organizations were included in the advisory board. The most important topic of the board was the rebuilding and redesign of the Linz Train Station. The advisory board met several times during the year. After the work on the Linz Train Station was finished in 2005, the advisory board was dissolved.[20] There is no passenger advisory board in the transport network. Customer complaints or inquiries are handled by the Upper Austrian Transport Network's customer centre. Special complaints or concerns are handled on an individual basis directly through the network's management board. According to the organization's data, all problems brought forth by

18 More information at http://www.verkehr.steiermark.at, accessed September 2007.

19 Contact info: http://www.land-oberoesterreich.gv.at/cps/rde/xchg/SID-3DCFCFC3-BF468F53/ooe/hs.xsl/ 20250_DEU_HTML.htm, accessed August 2007.

20 For an interim financial statement of the work see Höfler (2004).

customers are handled satisfactorily in this way. There has to date been no initiative for the institutionalization of citizen participation.

In the country *Vorarlberg*, there has been a transport concept in effect since 2006 called 'Mobil im Ländle'.[21] The concept replaced the 1992 document 'Verkehrsplanung Vorarlberg'. To provide further participation, the country instituted an Open Space Discussion Forum. The target group was youth between 13 and 20 years old who could take part in a discussion event.[22] There are no institutionalized participation possibilities for citizen organizations. On its website, the transport network offers the possibility of submitting a 'quality mail'.[23]

In *Tyrol*, the country government has a transport master plan. In one section, the plan deals with rail and public transport. Current developments for all means of transport are presented in annual transport reports.[24] There is no citizen participation in the country or the transport association in Tyrol. The intensity of informal cooperation depends on the occurrence of complaints. When customer complaints amass, some companies show a significant candidness to customer and transport networks. Between the Chamber of Labour and the stakeholders in the transport networks talks have taken place c. four times per year. However, since 2001, there has been no more noteworthy participation.[25] The Chamber of Labour has therefore been supporting the formation of a passenger advisory board in the Tyrol transport network. To date, the network has declined. Many contacts arise on the 'informal level'.

In *Carinthia*, there is a country development plan and a transport concept. The transport network of Carinthia coordinates the offering. There are no institutionalized participation possibilities. The association Fahrgast is included through the transport network in individual cases and according to the circumstance. An expansion of participation possibilities is not planned for the future.

In *Salzburg*, the current mobility concept[26] is effective between 2006 and 2015. There is no direct participation for citizens' organizations or user groups. The current operation serves the Salzburg network with 17 transport companies.

21 More information about this concept is presented at http://www.vorarlberg.at/vorarlberg/wirtschaft_verkehr/verkehr/verkehrspolitik/weitereinformationen/verkehrskonzeptvorarlberg.htm, accessed August 2007.

22 More information at http://www.rosinak.at/04_Highlights/News_21/04_Highlights/Highlightspercent20 Downloads/pressetext_pdf, accessed August 2007.

23 http://www.vmobil.at/06_Suche/index.html, accessed September 2007.

24 More information is available at http://www.tirol.gv.at/themen/verkehr/verkehrsplanung/, accessed September 2007.

25 Inquiry of the Tyrol Chamber of Labour.

26 More information at http://www.salzburg.gv.at/lmk2006, accessed September 2007.

There is a passenger advisory board in Salzburg, which resides with the city bus company. The passenger advisory board meets about twice annually and is informed about the company's new developments. Additionally, the transport operation instructs members with press communications, new brochures and manuals. Twenty-seven people actively participate. The Salzburg Chamber of Labour works with the passenger advisory board.

There is another passenger advisory board for the Flachgau-Bustakt, a region between the states Salzburg and Upper Austria. The passenger advisory board was formed in 2004 after connections began to deteriorate for commuters. In 2005 and 2006, there were several meetings, which finally lead to success. Since then, some connections have been directly routed to Salzburg. Presently the passenger advisory board meets when the occasion arises once or twice annually. Approximately eight people are active on the board.

5.1.4 Interest Representation and Lobby Groups

In Austria there are currently four inter-regional organizations that lobby for customer and passenger interests. These are the Chamber of Labour, Pro Bahn Österreich, FAHRGAST and the VCÖ.

5.1.4.1 Chamber of Labour[27] The Chamber of Labour (CL) is primarily a lobby group for employees (non-independently employed); it handles issues in consumer protection and gives legal advice. Their text about general consumer protection is as follows:[28]

> In 2003, 351,000 consumers sought advice and help from the CL's consumer advisory by telephone in writing or in person. In Vienna alone, there were 36,000 – with an increasing trend. … The CL's tests and practices from its advisory work show that there are still problems and that they are even increasing, for example in shopping, in financial services, contests or in the vacation industry. The position of consumers must be supported and they must be treated justly. For this reason, fair regulations are indispensable.

The organization, founded in 1920, is a public legal body. The CL has c. 2.6 million members. According to the Arbeiterkammergesetz (Chamber of Labour Law) of 1992, in basic principle all employed people belong to the organization. It finances its work through an allocation (a maximum 0.5 percent of social security contribution). The average total contribution is €9.12 per month. The Chamber of Labour is represented in the countries with its own organizations. It represents the interests of passengers with a focus on commuters. Non-members as well, however, (that is students, retirees) can

27 Website: http://www.arbeiterkammer.at, accessed September 2007.
28 http://www.arbeiterkammer.at/-192-IP-15867.html, accessed September 2007.

also make use of the organization. In a consultative paper addressing the EU Commission, the CL describes its work as follows[29]:

> In Austria, the Chamber of Labour (CL) is not only a representative of the interests of employed persons, but in recent years has been able to establish itself as a representative of passengers and commuters. The CL, collects all of the customers' concerns, desires, complaints about the rail companies (timetable, fares and reductions, delays, train cancellations, missing connections, additional connections, extension of service time, help with ex gratia agreements), bundles them and delivers them to the concerned transport companies or transport network organizations. We follow the solution to the problems and pursue them if no solution is presented.

The CL indicates problems in talks, and supports changes. It carries out active public work and conceptual work in the same degree. Its most important demands are short processing times for complaints, changed transportation conditions and improved ex gratia agreements for passengers.

5.1.4.2 Pro Bahn Österreich[30] Pro Bahn Österreich was founded on 5 January 2004. The organization originates from the initiative Wir sind die Bahn (We are the Train).[31] The organization thus orients itself by the organizations of the same name in Germany and Switzerland. Pro Bahn Österreich sees itself as a network of numerous local and regional initiatives. The number of its members cannot be conclusively stated. The 'circle of activists' consists of several hundred people. The total network consists of a few thousand people.[32] The organization works on a volunteer basis. Pro Bahn Österreich strives for a close cooperation with the European Passengers' Federation (EPF) [33] and corresponding organizations in Germany, Luxemburg and Switzerland.

The organization uses press and public work as a tool. The passenger representatives want to be better included on a regional and federal political level. Transport operations or transport networks include the organization only occasionally. Alongside public work, Pro Bahn Österreich uses directed lobbying work as a further tool. On political levels, round table talks or discussion forums are used. Pro Bahn Österreich supports a passenger advisory board with legal foundations.

---

29 http://europa.eu.int/comm/transport/rail/passenger/doc/bundeskammer-a.doc, accessed September 2007.

30 Website: http://www.pro-bahn-oesterreich.org, accessed September 2007.

31 Leading 'We are the Train' was the platform of the Salzburger Verkehrsinitiativen (Salzburg Transport Initiatives).

32 Numbers according to information from Peter Haibach, Pro Bahn Österreich.

33 Since 17 March 2004, Pro Bahn Österreich has been a member of the EPF.

5.1.4.3 FAHRGAST[34] Fahrgast[35] is a lobby group for public transport users. In 1985, a group of engaged private individuals founded the group. Fahrgast is a cross-party organization and is independent of any institution or company. The organization stands for a change of policy in transport politics. Public transport is meant to be privileged. Fahrgast supports shorter travel times through an acceleration programme for all lines. Furthermore, tighter departure intervals and mutually determined departures are meant to reduce waiting times and the need for transferring. Transport providers are meant to be appraised more strongly in terms of economic costs (environmental costs, accident costs). The offering for off-peak times (weekends, holidays, events) and at night should be improved. Fahrgast argues for improving organizations' participation.

5.1.4.4 Verkehrsclub Österreich (VCÖ)[36] The VCÖ was founded in 1988. The organization sees the goal of its work as safe, ecologically sustainable, and economically efficient mobility. The VCÖ represents particularly the interests of the weaker parties in transport. The organization counts among these children, the elderly, pedestrians, and cyclists. The organization represents passenger interests not directly but primarily through political lobbying work for improved basic conditions. In this way the organization advocates for a clear understanding of responsibilities between states and communities in public transport. Through passenger advisory boards, the customers are meant to take a stronger part than before in the planning of the offering.

Public transport is meant to be securely financed for the long-term. The VCÖ initiates occasional meetings and talks with stakeholders in transport politics and companies. The Verkehrsclub sees it as its mission to offer a basis for environmentally friendly decision-making in political, governmental and economic considerations of mobility. The organization has 15,000 members and operates its own research institute, which produces studies of, among other things, transport issues. The VCÖ supports an improved public transport, especially in rural regions.

5.1.5 Passenger Rights and Customer Service

5.1.5.1 Legal Foundation The legal foundation for claims in cases of delays or train cancellations is the Eisenbahnbeförderungsgesetz (EBG) (Rail Carriage Law) from 10 March 1988. In Section 24 it states:

34 Website: http://www.fahrgast.at, accessed September 2007.

35 The organization writes its name exclusively in capital letters in external communications.

36 Website: http://www.vcoe.at, accessed September 2007.

> *Train Delay and Cancellation*
> (1) If a train is delayed in departing or arriving or does not service a route in part or entirely, the traveller has no claim of damage.
> (2) If, due to a train delay, the connection to another train is missed, or if the train is missing in part or entirely from a route, the traveller can
> a) forego the continuation of his journey and request reimbursement according to Sections 29 and 42,
> b) forego continuation of his journey and demand to be brought to his original train station with the next available train with their baggage free of charge or
> c) continue his journey; in this case, the railway must transport him and his baggage without additional charge and as quickly as possible.
> (3) The railway must certificate, upon request, the missed connection or cancellation of a train. In cases of para. 2, b and c, the railway, as much as is possible, must extend the validity period of the travel ticket and validate it for the first class, for a train category with a higher fare or for a new mode of transport.
> (4) The railway must announce train delays of more than ten minutes, train cancellations and other operational disturbances.

There is no legal claim for compensation in cases of train delay or cancellation.

For buses in regular service, the new Kraftfahrliniengesetz (Road Services Law) has been effective since 1 January 2000.[37] The federal ministry responsible is authorized, based on the law, to introduce a regulation[38] about the safeguarding of standardized terms of transport (Section 46). Issues of liability and reimbursement of fare are to be regulated in the terms of transport. The liability issues addressed here are regulated in Section 47. This states:

> The company does not in any way guarantee the completion of trips according to the timetable. The company is particularly not liable for damages that occur due to train delays or absence.

This regulation is effective for all passengers in regular service transport. Hence for buses, similarly to trains, there is also no legal claim to compensation. According to Section 32, a transport company can waive the text in the

37 The full name of the law is the Federal Law Regarding Transportation of People on Routes with Motor Vehicles (Road Services Law) (Bundesgesetz über die linienmäßige Beförderung von Personen mit Kraftfahrzeugen (Kraftfahrliniengesetz (KflG)).

38 Verordnung des Bundesministers für Verkehr, Innovation und Technologie über die Allgemeinen Beförderungsbedingungen für den Kraftfahrlinienverkehr.

regulation. The 'special terms of transportation' are thus to be presented to the responsible authorities for approval.[39]

Another legal foundation is effective for street trams, underground trains and trolley buses. The Eisenbahnbeförderungsgesetz (Rail Transport Law) does not apply to them. Because they are not motor vehicles, the Kraftfahrliniengesetz (Motor Vehicles Law) does not apply to them either. They are subject to the Railway Act of 1957. According to Section 22 of the Eisenbahngesetz (Railway Act), the fare is comprised of the terms and the cost of the transportation. Every transport company is required to develop its fare system. The authorities must approve this if it does not conflict with public interest. In a company's terms of transportation, liability for delay or cancellation of a street tram is normally excluded. The terms of transportation for the Verkehrsverbund Ostregion, for example, are as follows:[40]

> *M. Cancelled departures, delays, cancellation and operational disturbances*
> The cancellation of departures or of connections, delayed departure or arrival of a vehicle, as well as operational disturbances, and operational interruptions of all kinds, and shortage of space do not justify a claim to compensation as long as the damages are not caused by grossly negligent conduct on the part of the transport company's employees. If the passenger is not entitled to compensation in the respective case, neither a reimbursement of the fare nor free transportation of the passenger is to take place. In cases of delay and cancellation as well as lack of space, however, the transport company is required to the best of their ability to provide a means of transport along the same or another route to enable the passenger to reach his destination with the least possible delay, and with no additional cost.

Different legal foundations are applied depending on the means of transport. The result is, however, mostly the same. In cases of delay or cancellation, there is no claim for damages. Alongside legal foundations, passengers must make themselves aware of the transport associations' terms of carriage, as long as they have been travelling with a means of transport belonging to the association. The terms of carriage allow at least a claim for onward travel to be derived.

5.1.5.2 Ex gratia *agreements* The ÖBB instituted a passenger charter in December 2004. According to this, in cases of a delay of 60 minutes or more in national transport, passengers are reimbursed 20 per cent of the fare. For

---

39 In the context of this work, no case is reported on in which a transport company extended the terms of transportation to special terms of transportation in order to improve the legal position of a passenger in a case of delay or cancellation of a train.

40 http://www.vor.at/bilder/d1/BefoerderungsbedingungenA41.pdf, accessed September 2007.

international transport, the charter covers delays starting at 120 minutes. The reimbursement is limited for certain tickets or incidences such as cases of violence or strike. The regulations are oriented towards the recommendations of the Community of European Railways and Infrastructure Companies (CER). Local transport by rail is only included in the charter in situations involving multimodal travel which includes at least one long-distance train.

In individual cases, all transport companies offer *ex gratia* agreements. This includes the coverage of taxi costs, if the passenger has missed the last connection. All further regulations such as redemption of tickets or overnight accommodation are at the discretion of the transport company. There are presently no further *ex gratia* agreements such as mobility guarantees, voluntary company commitments or special regulations for associations.

5.1.5.3 Arbitration boards Since January 2007 there has been an arbitration board in Austria.[41] The arbitration board is located at the Österreichische Schienen Control GmbH (SCG) (Austrian Rail Control). The SCG is supervised by the BMVIT. Dissatisfied passengers must first register a complaint with the appropriate transport company. If customer and company cannot come to an understanding about a particular circumstance, the arbitration board makes a settlement recommendation for both parties. The legal foundation is the Eisenbahngesetz (Railway Act). In this law, Section 78 a and b were added in 2006. A special directive, which was decreed on the basis of the law, regulates further conditions for the arbiters' work. The institution is not responsible for street trams or bus transport.

In the first half of 2007, the arbitration board received around 60 complaints. Most complaints were against the largest company in rail transport, the ÖBB. Three complaint groups were especially common:[42]

1. ticket machines, criticism of the timetables, fare structure;
2. poor information, accessibility, homepage;
3. delays, congestion, cancellation of trains.

The actual type of complaint varied. Defective ticket machines, the lack of a night train to other European cities, gaps in the timing of local transport, poor connections or long travel times were reasons for passenger aggravation. On the one hand, unfriendly personnel were criticized, on the other, friendly personnel were expressly commended. Information, especially in cases of disturbances in train service, could be better, according to customer opinion. The ÖBB's Internet offering, too, is not always what the customers would desire.

41 Website: http://www.scg.gv.at, accessed September 2007.

42 Inquiry with the SCG.

It is critical for the SCG to examine the ÖBB's compensation policy. The SCG considers it upgradeable. The arbitrators see the ÖBB's passenger charter as critical as well. The maximum reimbursement of 20 per cent is particularly unsatisfactory. The confusing fare determinations between the ÖBB and the transport associations could be simpler. The arbitration board now aims to increase its visibility and cooperate with organizations in protecting users. To date a majority of the complaints have come through the Ministry of Transport.

The arbitration board must produce an annual report. In the report, the essential problems must be presented as they are determined through the processing of complaints. The arbiters' work is meant to generate suggestions for improvement. The details are determined in the directives for the arbitration board's work.

5.1.6 Conclusion: Need for Action in the Development of User Protection and Participation

Participation possibilities for passengers and the representatives of their interests in Austria are few. An occasional inclusion or circumstance-based consultation of organizations is certainly not sufficient to create lasting and trust-based cooperation between those involved. In some associations, participation is declined or only carried out on the basis of the transport contract, as in Vienna. There are no legal claims for participation. This is astonishing considering that the legal guidelines have only been in place for a few years. In legislative procedures, stronger citizen participation could have been introduced without problems.

The situation with passengers' and customers' rights also bears improvement. The companies show themselves to be accommodating in individual cases. There are practically no legal claims or claims based on general terms of carriage. There is, however, something to aid passengers: the passenger advisory board at the ÖBB, the passenger advisory board at Wiener Linien and some local and/or regional activities are a step toward more openness. The comprehensive participatory practice for the Transport Masterplan in Vienna, too, can act as a role model. Since 2007, there has been a new arbitration board – however, only for rail transport. Nonetheless, this board's required annual report increases transparency and candidness towards the passengers.

Overall, the situation in Austria remains unsatisfactory. Marginal participation possibilities and passenger rights, little transparency – for example, in relation to customer complaints – and the service providers' strong reservations about passengers – could all be significantly improved. In this way, there is ample room for all of those involved to improve the current situation of passengers.

5.2 The Czech Republic

Holger Jansen[43]

5.2.1 Political and Geographical Context

The Czech Republic is a parliamentary democracy with c. 10.3 million residents (as of 2006) and a land area of 79,000 km^2. The largest city and also the capital is Prague (Praha in Czech) with c. 1.2 million residents (as of 2006). Other large cities are Brno, Ostrava and Pilsen (Plzen). The country has belonged to the European Union since 1 May 2004

The Parliament consists of two chambers: the Chamber of Deputies (lower house) and the Senate (upper house). The head of government is the Prime Minister. Since the structural reform in 2000, the Czech Republic has consisted of 14 regions each with their own administration and representative committee.

The representative committees decide above all matters of economic development in their regions, the administration of hospitals, schools, the maintenance of monuments and road construction. The legal foundation is the 2000 law regarding the composition of the regions. The representative committees also deal with the planning of public transport. The law states:[44]

> Section 35 Authorities of the Representative Committee
> (1) ...
> (2) The Representative Committee is held to
> (a)–(c) ...
> (c) issue generally binding area declarations,
> (d) coordinate the development of the area's land, to adopt programs for the development of the area's land according to particular laws, to ensure their realization and to monitor their abidance
> (e) to adopt territorial planning documents for the area and to announce their obligatory portions through generally binding regional declarations,
> (f) to choose representatives for the area in regional councils of regional partnerships,
> (g) to determine the scope of the ground transport services for the area
> (h)–(s) ...

In the administrative arrangement, under regions are districts. The smallest independent administrative units are city and municipality.

---

43 I would like to thank Štěpánka Doleželová, Nicolas Bach and the Czech Ministry of Transport for assistance in preparing this section.

44 http://www.collegium-carolinum.de/doku/vdok/kreis-cr.htm,accessedSeptember 2007. In the following text, the Czech word 'kraj' is translated with 'region'.

5.2.2 The Transport Market and the Organization of PT

The Ministry of Transport (Ministerstvo Dopravy)[45] is responsible for transport in the country. The Ministry is organized into specialized departments for roads, waterways, railways and public transport and is responsible for transport nationwide. Regional transport is organized by the regions, districts and municipalities. Inner-city transport is coordinated by the respective city. Individual regions have founded transport associations in order to organize a cohesive offering.

The Transport Research Centre is responsible for research duties. It is answerable to the Transport Ministry. The Institute addresses topics of public means of transport, statistics, safety and other topics having to do with transport.[46]

The most important legal foundation for rail transport is railway statute 266/1994 from the 14 December 1994.[47] This contains the ordinances for building and operation of railways and street trams.[48] Statute 111/1994 is effective for public passenger transport (PT). Further regulations[49] contain ordinances about timetables and regulations for carriage. In the statutes 128/2000 and 129/2000 are ordinances regarding the financing of LPT through the regions.

In Prague, Brno and Ostrava, there are transport associations with organizational functions.[50] In the other regions, there are federations, but political and company-related aspects are to date not clearly delineated. Politics and business overlap in terms of the offering. There are departments responsible for transport on the political level in the regions.

Medium-term transport and service planning occurs through the respective political levels. There is a Program for District Development in the Area of

---

45 http://www.mdcr.cz/en/, accessed September 2007. Much of the information is available in English.

46 More information at http://www.cdv.cz/english/index.htm, accessed September 2007.

47 Law no. 266/1994 of 14 December 1994 regarding railways. Consolidated version from 1 July 2006 available at http://www.mdcr.cz/NR/rdonlyres/A047D156-F3C5-4B4A-ADC8-BF930719244A/0/z26694k _1_7_2006EN.rtf, in English, accessed September 2007. Further legal foundations for liability in rail transport are the road transport law, the Civil Code, and the transportation regulation for public rail transport. For international travel, the COTIF accord is also effective.

48 The law does not apply to industrial trains, trains used in mining, movable trains and ski lifts.

49 Details in Czech at http://www.mdcr.cz/cs/Legislativa/Legislativa/Legislativa_CR_drazni/, accessed September 2007.

50 More information can be found at http://www.ropid.cz, http://www.kordis.cz and http://www.kodis.cz, accessed September 2007.

Transport or General Plan for Transport Development.[51] The respective transport departments are responsible for this. The legal foundations for the plans are statutory guidelines, such as the railway statute, the municipal statute, the district statute, the statute regarding the capital, Prague, and other statutes that concern transport development. Financing is predominantly through public means.[52]

The part of the service that is not covered by income from fares is financed by the national, district or municipal governments. In 2000, the national government transferred financial responsibility for local transport to the regions. In 2005, the government subsidized railway transport with CZK2.62 billion (c. €92 million).[53] The regions used CZK7.166 billion for rail transport (c. €253 million). In total, the political agents subsidized public local and long-distance transport with c. CZK11.5 billion (c. €407 million).[54] Fare income covers 42 per cent of costs in city transport in Prague, in Brno 54 and in Ostrava 35 per cent.[55]

Between 2000 and 2005 the trains lost about 2.5 per cent of their passengers. Public bus transport in the same period lost 12 per cent of its users. Only air transport increased – by 3.5 million passengers in 2000 and 6.3 million passengers in 2005 (an additional 80 per cent). The passenger count on waterways – c. 1 million – only played a role at vacation or off-peak times.

5.2.2.1 Local transport In 2005, statistics for all public means of transport showed around 2,843 billion passenger journeys. Approximately 180 million were on trains and about 386 million on buses: 2.27 billion customers used public means of transport in cities. The market share of public means of transport dropped by approximately 2.5 per cent in the period between 2000 and 2005.

The largest municipal transport company is the transport operation in Prague. Approximately 2,000 vehicles transport passengers in the city and region daily. The just under 12,600 employees serve in the region of 1.1 billion people annually (as of 2006).[56] The operation was reformed in 1991 and since then has operated as a corporation. The city of Prague is the owner. The company links to the hundred-year old tradition of its predecessor. Gradually, an integration took place in the first half of the 1990s of the individual lines in a common fare system. Presently, some rail routes in the ČD are being

---

51 Inquiry with the Ministry of Transport.

52 Inquiry with the Ministry of Transport.

53 In January 2008, the exchange rate was 1 CZK to €0.0380.

54 *Statistical Transport Yearbook*, http://www.sydos.cz/en/yearbook-2005/index.html, accessed September 2007.

55 Sdruzeni dopravnich podniku, *Yearbook 2005*, http://www.sdp-cr.cz/archiv/vz2005_v2.pdf, page 28. Accessed September 2007.

56 http://www.dp-praha.cz, accessed 2007.

integrated into this. Prague Integrated Transport (Pražská integrovaná doprava (PID)) offers a unified fare for all means of transport. In local city transport in Prague, there is the underground train, street trams, buses and a funicular railway.[57]

Viamont, a train company that belongs to the French Veolia Group, serves two routes in North Bohemia. The train Connex Morava, which originated with the same group, serves a rail route and a bus line. In total, Veolia Transport has seven subsidiary companies in the Czech Republic.[58] According to the company, they are the largest private provider of transport services in the country and earned a gross income of CZK2.23 billion in 2006 (€79 million).

The company Jindřichohradecké místní dráhy (JHMD) operates two narrow gauge lines. In border regions there is cooperation with transport operations in neighbouring countries. An example is the route Liberec-Zittau. Alongside the railways there is a long-distance bus system. On certain routes the buses travel in close succession – for example, between Liberec-Prag – some every 30 minutes, on other routes only once a day.[59]

5.2.2.2 Long-distance rail transport The rail network in the Czech Republic comprises 9,614 km.[60] In 2004, 46 different railway companies were active in the Czech Republic. This number refers, however, to small companies as well which are only active in the transport of goods. The largest railway company and market leader in long-distance transport is Česke dráhy (ČD). Since railway reform in 2003, ČD has been operated as a privatized corporation. The usage of railways as part of the country's transport infrastructure is regulated by contract with the ČD. The company transported approximately 170 million passengers in 2003.[61]

5.2.3 Transport Planning and Participation

Every provider of a regular service requires a concession. This is true for rail as well as public passenger transport (PT). Companies must register the concession with the department responsible for public transport. For this registration, the administration requires information about the planned lines, vehicles, and

57 Extensive information in English is available at http://http://www.dpp.cz/index.php?q=en/prague-integrated-transport-system accessed January 2008.

58 More information at http://www.connexmorava.cz, accessed September 2007.

59 http://www.nahverkehr-ostsachsen.de/html/body_fernbus.html, accessed September 2007.

60 *Transport Statistical Yearbook* 2005, Transport Research Centre, http://www.cdv.cz/english/index.htm, accessed September 2007. There is also an additional c. 400 km route for trolley coaches, 364 km for street trams and 50 km for metro trains.

61 *ČD Statistical Yearbook* 2003, http://www.cd.cz/static/sr/CD_rocenka2003_ang.pdf, accessed September 2007.

planned offering from the applicant. The applicant must also indicate the offering's usage possibilities for people with limited mobility. Citizens cannot directly participate in the determination of whether a concession is granted or denied. The decision is carried out by the administration.

Citizens do, however, have the possibility to submit their desires and recommendations to the affiliates responsible for the award of concessions or to the city council. There are bodies connected to the city or region's population, including the city council and special affiliates for transport, which are occupied by representatives from political parties. In larger cities, experts from the transport industry are on the boards alongside politicians. The affiliates are involved in the process of awarding concessions and determine the long-term goals for transport in the city. The final decision for the award of a licence is made by the administration. The exact form of participation is determined by each city or region individually. There is no legal foundation for citizen participation.

In other cities or regions, participation can take place in other ways that that presented here. There are no direct participation possibilities for passengers in transport companies or transport associations.

The central authority in Brno is the transport organization Kordis.[62] In rail transport, citizens can communicate their desires and concerns directly to the company through České dráhy. The company forwards the concerns to Kordis. The cities and municipalities involved discuss the concerns. In this process, the participants seek to find an agreement about the recommendations. This practice takes place in writing. The participants call a meeting in cases of serious problems. This happens particularly when no agreement can be reached in written correspondence. Approximately 200 cities and municipalities belong to the region that Kordis serves.

In regional bus transport, distinctions have to be made between those bus lines integrated in the association and those not. Kordis addresses all concerns for bus transport in the association and tries to accommodate the citizens' wishes. The district's transport department is directly responsible for all other bus lines. Desires and concerns must be processed there.

In Brno's intra-city transport, the city council, the local transport provider DPMB and Kordis try to act in cooperation with the individual city districts to address customer desires.

Middle-term goals are described in the plan for national transport policy. The current plan is effective between 2005 and 2013. In this plan, goals and focuses are identified for all means of transport. According to the customers of public transport, quality and service particularly should be improved. The existing passenger rights are to be revised to address, among other things, quality criteria and determinations for the payment of reimbursements.

---

62 Website: http://www.kordis.cz, accessed September 2007.

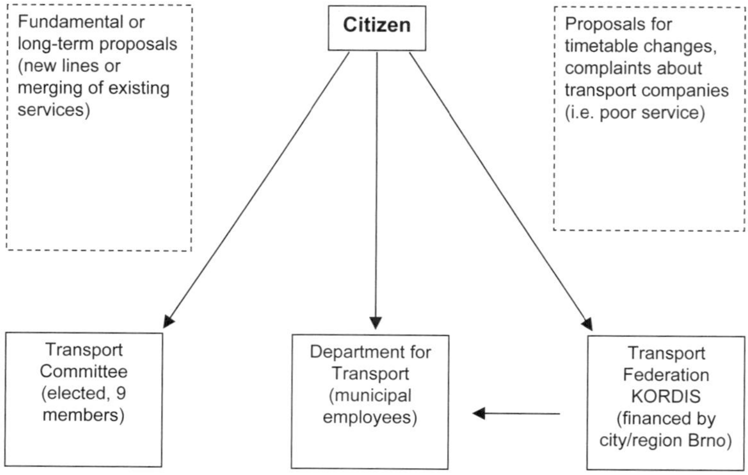

Figure 5.1 Citizen participation in public transport in Brno

Source: author's depiction based on information from Štěpanká Doleželová.

All public means of transport are to be accessible for people with limited mobility.

5.2.4 Interest Representation and Lobby Groups

5.2.4.1 Transport association of the Czech Republic[63] The transport association was founded in 1989 as an amalgamation and interest representation group of numerous companies in the transport industry. Since 1994 the transport association has been a member of the industry's Czech umbrella organization. The transport association is the largest single member of the umbrella organization with 143 companies and 140,000 employees. The industry association creates a platform for 25 associations. The industry association's member companies employ 1.3 million people.

The transport association has sections for all means of transport. Particularly relevant for public transport are the sections for rail transport and local public passenger transport (LPT). In local public transport particularly, the interests of the public transport service are meant to be represented in the city and in relation to the region and self-administration. Their attention is directed primarily towards:

- creating favourable general regulations for financing public city transport;

63 Svaz dopravy České republiky' (SD ČR), http://www.svazdopravy.cz, accessed September 2007.

- producing arguments for LPT in order to ensure financing from the national budget;
- demanding unified fares for integrated city and regional transport;
- processing claims for the transport policy of the Czech republic, especially those concerning LPT;
- doing political lobby work in order to promote a unified LPT system to the parliament and government.

For the period until 2010, the section sees is main strategic goals as:

- improving quality in LPT in consideration of environmental aspects;
- stabilising the transport ratio between LPT and individual motor transport in cities. Goal: LPT's market share should not sink below 55 percent of the entire transport service in 2010;
- LPT should be clearly favoured in city transport;
- LPT's vehicle fleet should become more attractive through new and modernized vehicles. Public transport's competitiveness must be increased as compared to individual motor transport.
- The Czech Republic's general legal regulations must be normalized with those of the EU especially in areas with region involvement. Of particular note here are local public transport's financing, development and renovation.

5.2.4.2 Association of Train Customers (Sdružení železničních zákazníků)[64] The Association of Train Customers is a member of the European Passengers' Federation (EPF).[65] The association sees itself as a group that represents train customers' interests. That refers to transport of passengers as well as of goods. The organization advises and informs its members about current developments. It was funded in 2002.

5.2.4.3 Union for Public Transport (Spolecnost pro verejnou dopravu)[66] The Union for Public Transport, founded in 2001, is active throughout the Czech Republic. The focus of its work is urban transport. The spectrum of membership ranges from transport specialists to people who are interested in the development of urban transport. The union advocates an improved offering in public urban transport and calls attention to the impact of motorized individual transport (MIT) in cities. Programmes are meant to be developed with the cities that address these problems. LPT is meant to be strengthened with exhibitions or studies. Furthermore, the union campaigns for more modern vehicles and more appealing fares.

64 Website: http://www.asymetric.cz/sdruzeni, accessed September 2007.
65 Website: http://www.epf.eu, accessed September 2007.
66 Webseite: http://www.spvd.cz, zuletzt September 2007.

5.2.4.4 Czech Consumer Protection Union (Sdružení obrany spotřebitelů České republiky)[67] The Czech Consumer Protection Union was founded in 1993. The union advises consumers about their rights, publishes trade magazines and brochures. The union is non-profit and independent of government or other region bodies. The consumer protectors advocate for safer products, collaborate on drafting of bills for better consumer protection and strive to effectively represent customers' interests. The union is a member of the worldwide organization Consumers International and of the European Consumers' Union (BEUC). In seminars, workshops and congresses, the representatives exchange their experiences in consumer protection in times of market globalization on an international level.

5.2.4.5 The Transport Association SDP (Sdružení dopravních podniků ČR)[68] The SDP was founded in 2002. It supports cooperation and information exchange between urban transport providers. This is meant to improve the organization, the service and the cost-effectiveness of the operation. The members include 19 large urban transport operations and 109 others. This membership represents all major Czech cities. Through specialty work groups, the association supports the work of transport services in the following areas:

- operation of local transport;
- street trams, trolley coaches and omnibuses;
- cost-effectiveness;
- technical issues;
- personnel issues and passenger checks.

5.2.5 Passenger Rights and Customer Service

The most important legal foundations for train passengers is the railway statute, which must be the basis of companies' terms of carriage. Claims in cases of delay or train cancellation do not exist in national transport. In long-distance transport, when trains are delayed more than 60 minutes ČD customers get back additional charges for long-distance trains (Intercity/Eurocity).[69]

The ČD has instituted the Railway Passenger Transport Charter. The umbrella organization of European trains (CER) developed this document.[70] As long as the company is at fault for a delay, the passenger must be appropriately assisted. This refers to continued transport or overnight accommodation as long

67 Website: http://www.spotrebitele.info/lang, accessed September 2007.

68 Website: http://www.sdp-cr.cz/vz.asp, accessed September 2007.

69 The legal basis for this is in the ČD's terms of carriage under numeral 307, http://www.cd.cz/static/sppo/sppo.html, accessed September 2007.

70 Extensive text at http://www.cer.be, accessed September 2007.

as the destination was not reached. The charter also contains determinations for regular consultations with customers or representatives of their interests.

In the terms of carriage for the remaining forms of local transport, companies can create regulations above and beyond the statutes for the benefit of the customer. There has been nothing like this until now. Some transport operations give quality promises on a voluntary basis. The transport company Dopravní Podnik from Děčín[71] requires itself to be punctual. The goal is generally maintained and contains no requirement of reimbursement payments. The company Bus Sleszsko sro,[72] which is a member of the Connex Group, has promised new buses, passenger support in cases of cancellation, friendly personnel and defines punctuality as its goal. ČSAD in Semily has promised customers modern buses and well-trained drivers.

The Czech transport association SDP creates standards for service quality. On a voluntary basis, parameters are meant to be determined for punctuality, information, personnel friendliness, availability of information and tickets as well as cleanliness of vehicles.

In cases of complaints, passengers can address the company directly. The customer has a maximum of six months for this. There are no legal determinations for the time period in which customer complaints must be processed. For members of SDP, complaints must be processed within 30 days.

Details about the legal relationship between passenger and company are determined in the terms of carriage. In article 37, the railway statute requires companies to produce terms of carriage. The text is:

> The conditions in which a passenger and his baggage are to be transported by means of rails, including transport of goods, must be determined in terms of carriage.[73]

After this, the points to be regulated in the terms of carriage are extensively described. Among others, this includes entries about contractual partners, essential information on the ticket, modes of payment, seating plan requirements, terms of carriage for children and people with limited mobility and return of tickets.

In 2000, the Ministry of Transport passed an ordinance (175/2000) concerning the terms of carriage for public railway transport and passenger transport on the street. The legal basis for the ordinance is the railway law, article 37. Passengers' rights are regulated in this ordinance. In Czech, the terms of carriage are called Smluvni prepravni podminky.

Passengers only have legal claims in the Czech Republic in civil law in cases of traveller death and injury. In cases of property damage, travellers are

71 More information at http://www.dpmdas.cz, accessed September 2007.

72 More information at unter http://www.connex.info, accessed September 2007.

73 Quotation from the Ministry of Transport's homepage (English version).

compensated at the rate of CZK 100 per kg of baggage, to a maximum of 1,500CZK (c. €53). There are no more extensive legal claims. The customer is thus dependent on the company's voluntary consent.

There is no arbitration board for alternative dispute resolution.

5.2.6 Conclusion: Changing Transport Systems – Are Opportunities for Citizen Participation being Used?

The situation of public transport in the Czech Republic is at present influenced by constantly changing basic conditions. After the political upheavals and changes of the 1990s, individual transport has been consistently increasing. Public means of transport had to more or less accept significant losses while other means of transport were able to stabilize themselves extensively.

Stabilizing the market share is one of the most important goals in national transport politics. With new vehicles, improved accessibility and improved information, public transport should be stronger by 2013, according to current plans.

Another focus is legal basic conditions, which must be converted through the Czech Republic's membership in the EU. In this, the separation between corporate and political responsibility in local and regional transport becomes especially apparent. Transport associations are offering new opportunities. One fare, one ticket and one determined offering are advantages that are successfully practiced in other regions of Europe. This creates new opportunities for active citizen participation, for example, with a passenger advisory board in the transport association.

Citizen participation in the Czech Republic's transport sector is very limited. Citizens can communicate their concerns or wishes to local institutions. There are no special interest groups such as passenger or consumer advisory boards. This presents a potential for improvement on the side of the transport companies or federations.

In passengers' rights, the customer is currently dependent on the company's voluntary consent. There are legal claims in only a very few exceptional cases. New momentum in this area will only come, on the one hand, through the EU, which is currently redefining the conditions for passengers' rights in train transport. On the other hand, voluntary promises from the companies offer an initial point for continued dialogue between consumer and provider associations.

Associations are divided into general consumer associations and specialized transport associations. They are represented with different points of focus. They provide a good initiation point for cooperative work in strengthening local public transport. There is currently no independent arbitration board for passenger grievances. This is a point where there is a need for action in order to further strengthen the customer's position.

5.3 Sweden

Holger Jansen[74]

5.3.1 Political and Geographical Context

Sweden is a constitutional monarchy and has 9.12 million residents (April 2007). Approximately 783,000 people live in the capital, Stockholm, c. 1.92 million in the county. Other major cities include Gothenburg with 490,000 residents and Malmö with 258,000 residents. The land area measures 450,000 km^2. This makes Sweden the third largest country in Western Europe. The country has been a member of the European Union since 1995. The head of region is the monarch. The Swedish Parliament is called the Riksdag in Swedish. The country is divided into 21 counties (*län*). Political work in the counties is coordinated by County Administrative Boards. The smallest administrative divisions are the 290 cities and municipalities. They are responsible for, among other things, school systems, social services, care of the elderly, support of the handicapped, as well as land use and building regulation. Parliament is responsible for legislation.[75]

5.3.2 Transport Market and Organization of PT

The Ministry for Industry, Labour and Communication coordinates the planning and design of the transport network. Above all, the Ministry leads long-term planning projects in the transport area. This includes concepts for good accessibility to public transport, high quality, safety, environmental friendliness, aspects to strengthen the regional structure, and consideration of the interests of men and women. The goal of this nationwide transport service is, on the one hand, efficiency, on the other it is to ensure sustainability.[76] Means of transport are to be better networked. Public transport is to be made more effective and cost efficient through stronger competition. On the county level, transport service should plan and design as well as be aware of users' interests. According to a resolution in Parliament in 1998, all of those involved in transport policy should be more aware of aspects concerning sustainability.

Three organizations play an important role in public transport: Rikstrafiken, Banverket, Järnvägsstyrelsen and Vägverket. Rikstrafiken was founded in 1999. This organization's task is to coordinate transport on a national level. Rikstrafiken contracts out the transport that is not provided by commercial

---

74 I should like to thank Kurt Holtgren of Resenärsforum and Ola Svenonius for their support in creating this country report.

75 The counties have no legislative competence.

76 Ministry for Industry, Labour and Communication, http://www.sweden.gov.se/sb/d/ 2156/a/20469, accessed September 2007.

means and negotiates contracts with transport companies.[77] Rikstrafiken's work concerns all modes of transport.

Banverket has been responsible for the rail network since the separation of network and transport in 1988. In 1999 Banverket became responsible for the production of timetables as well.[78] Banverket counts its most important task as the creation of a transport system in the entire country that is environmentally friendly and that is in accordance with the fundaments of sustainable development.[79] Public transport is to be accessible for all customer groups and to take aspects of regional development into consideration.[80] Approximately 6,500 employees work for Banverket. Contact to passenger and consumer associations should, according to the organization, be expanded in the future. Based on a plan passed by the government in February 2004, Banverket is intended to invest SEK108 billion (€10.9 million) in infrastructure.

Järnvägsstyrelsen[81] is the regulatory body according to EU law. The office is responsible for safety in rail transport and discrimination-free access to the network. The organization must provide for fair competition between individual companies. The organization, with its headquarters in Borlänge, has existed since 1 July 2004.

Vägverket [82] is first of all responsible for building and maintenance of road networks. The organization is, however, also responsible for issues of public transport on roadways, of pedestrian and bicycle traffic and of issues concerning 'soft topics'.[83]

Counties and municipalities have been the transport authorities for local passenger transport (LPT) since 1978. For this reason they founded Public Transport Authorities (Trafikhuvudman). There are currently 21 Public Transport Authorities in Sweden.[84] They are responsible for fares, tender procedure and assignment of concessions. The Authorities also plan the services, coordinate the individual companies and ensure quality and cost-

---

77 More extensive presentations can be found at http://www.rikstrafiken.se, accessed September 2007.

78 Banverket aims to improve the flow of information between operators and Banverket in cases of service disturbance.

79 Website: http://www.banverket.se, accessed September 2007.

80 More information about Banverket's goals and tasks at http://http://www.banverket.se/en-gb/Amnen/About-Banverket/Duties-and-organisation/Goals.aspx, accessed September 2007.

81 Webseite http://www.jvs.se, accessed September 2007.

82 Webseite http://www.vv.se, accessed September 2007.

83 This includes mobility services for people with limited mobility.

84 All Authorities are members of the association SLTF (Svenska Lokaltrafikföreningen). The association is a conglomerate of regional and local transport operations. The members include all 21 Public Transport Authorities as well as 80 other transport companies. Eight employees coordinate SLTF's work. Website: http://www.sltf.se, accessed September 2007.

efficiency. The Authorities inform customers and link the LPT service with other transport providers.

Local public transport is financed on a national, county and municipal level. In 2005, SEK23.8 billion (€2.58 billion)[85] was used for this purpose. 58 percent (SEK13.7 billion – €1.48 billion) were covered by income. The municipalities spent SEK2.83 billion (€307 million) for public transport, the counties SEK7.35 billion (€797 million). Further contribution margins consist of c. SEK257 million (€27.9 million).[86]

Approximately 60 per cent of roads are maintained for passenger motor vehicles. Thirty per cent are dedicated to pedestrians or bicycles. The remaining 10 per cent are dedicated to public means of transport.[87] Every citizen used public transport an average of 124 times in the year 2005. Not including the Stockholm metropolitan area, this figure is just 70 journeys per person and year. In 2005, 1.121 billion passengers used public transport (499 excluding Stockholm). Between 1999 and 2005, the passenger count rose by 7 per cent. For long-distance transport, the train achieved a market share of 15 per cent and the bus 5 per cent. In local transport, the train achieved a market share of 4 per cent and the bus 9 per cent.[88] Current figures are published by the Statistics Bureau[89] in various reports about rail transport and public transport.

The rail network comprises a total of c. 17,000 km.[90] The networks in the public domain comprise almost 14,500 km. Other operators (private trains, quayside railways) account for c. 2,500 km. The largest provider of rail transport is Statens Järnvägar (SJ). In 1998, the government decided that SJ was to be made into corporations. Since 2001, SJ has been divided into six independent corporations. They are all publicly owned. In terms of passenger-km, SJ had a market share of 98 per cent in 1998.[91] In 2003 its market share was c. 75 per cent.[92] Besides SJ, seven other rail companies have a licence for passenger transport, ten for transport of goods.

The national bus market is shaped by four major providers. These are Swebus Express AB, Säfflebussen AB, Svenska Buss and Ybuss with a total market share of c. 90 per cent. The largest local public transport operator is 'Göteborgs Spårvagnar' with c. 2,500 employees and 560 buses. The company

85 As an exchange rate, €1:SEK9.22 has been chosen (May 2007).

86 http://www.sika-institute.se/Doclib/2006/ss_2006_24.pdf, accessed September 2007.

87 Motives for and extent of travel are not more closely examined here.

88 These figures comprise the national average. In the Stockholm metropolitan area, modes of public transport have significantly larger market shares.

89 For rail transport see http://www.sika-institute.se/Doclib/2006/ss_2006_19.pdf, accessed September 2007.

90 Statens Institut för Kommunikatinons Analys (SIKA) Annual Report (2003), p. 25 in English, http://www.sika-institute.se, accessed September 2007.

91 SIKA: Järnvägar (1998).

92 Reference parameter: passenger-km, Banverket (2003).

operates eleven street tramlines. Besides the major companies, there are many smaller bus companies. For some long-distance transport connections, there is bus-line transport parallel to rail routes (for example, between Stockholm and Gothenburg). The bus market has been deregulated since 1989.

5.3.3 Transport Planning and Participation

In planning, a distinction must be made between short- and long-term planning. Long-term plans mostly apply to the next five to ten years. The Stockholm metropolitan area is currently planning for the next 15 years. The first strategic plan for the Stockholm metropolitan area Storstockholms Lokaltrafik (SL) was created in 1990. The current plan was created in 1999 and is currently being revised. There is no legally required participation of a citizens' association or interest representation group; participation occurs on a voluntary basis.[93]

Short-term plans refer to the coming transport year. The transport association creates the plans and submits them to the county administrative board. The county administrative board[94] decides their implementation. The transport association creates a report for the county administrative board about goals determined in previously created plans.

The transport authority can include consumer and passenger associations in the discussion process. There is no legal claim to this inclusion. There is regular dialogue with associations who represent people with limited mobility.[95] For the Stockholm metropolitan area, there has been a voluntary agreement between SL and the associations since 1981 to maintain a regular dialogue. This takes place four to five times per year. Work groups are organized for special projects.

In the Stockholm metropolitan area, SL serves a transport area of approximately 1.9 million residents. On workdays, c. 670,000 passengers use the service.[96] The transport authority aims to improve quality, safety, accessibility for people with limited mobility and the environmental friendliness of transport in the coming years. The percentage of satisfied customers rose between 2000 and 2006 from 56 to 64 per cent.[97] SL surveys passengers with statements such as, for example 'the bus is always clean'. The customers then answer from 'fully agree = 7' to 'do not agree at all = 1'. 'Undecided' is indicated with '4'. Passengers with values 5–7 are considered 'satisfied'. Values 1–3 are considered

93 Inquiry with the SL.

94 In the Schonen metropolitan area (Skåne län, Malmö) and West Gothenland (Västra Götalands län, Göteborg) decision-making occurs through the regional parliament.

95 Inquiry with the Ministry of Transport.

96 SL annual report 2006, p. 1.

97 SL Jahresbericht 2006, p. 6.

'unsatisfied'. The goal is to reach a count of c. 80 per cent satisfied customers in all modes of transport by 2010. Over 90 per cent of trips, with the exception of regional trains, were punctual. An extensive account is available in the SL annual report. Particularly in cases of delay, SL aims to improve information to travellers.

Service planning takes place through SL in close cooperation with the 26 city districts. SL analyses the customers' desires with various instruments. There is daily dialogue between the management and customer service. 50,000 passengers gave their suggestions, ideas and complaints to SL. These addressed, among other things, bus lines, safety on train platforms, and soiled facilities.[98] The owner of SL is the county of Stockholm.

5.3.4 Interest Representation and Lobby Groups

5.3.4.1 Järnvägsfrämjandet[99] 'Järnvägsfrämjandet' is an independent passenger organization. The association is led by a head managing board and four regional divisions with their own boards of directors. Each region has its own regional representative. The association works on a volunteer basis and is independent of parties, transport associations or transport companies. A few transport associations are among their c. 900 members. The passenger representatives have two main goals: one is to strengthen the railways and highlight the positive qualities of trains. The other is to increase the usage of railways in the transport of goods and passengers. Rail transport is to become an alternative for the environmental damage and energy waste of road and air travel.[100]

Järnvägsfrämjandet cooperates particularly with public transport associations. Other contacts exist to politicians, transport companies and other interest groups. The association's main task is active media and lobby work. The members' magazine and informational brochures have a high priority as well. They have no legal claim to participation in the planning done by companies or transport authorities. The organization, however, does write to these bodies or give their indications in talks. Stakeholders in Järnvägsfrämjandet do not value participation possibilities in planning processes very highly. There are, however, exceptions: this is how the regional division east was able to avert poor service in regional rail transport. Once it was possible to prevent a shutdown of regional services. In these cases, association representatives intervened in the planning phase. The association is positive about its cooperation with Banverket to date. Banverket makes preliminary studies and general research

98 SL Jahresbericht 2003, p. 10.

99 Website: http://www.jarnvagsframjandet.se/index.htm, accessed September 2007.

100 Details at http://www.jarnvagsframjandet.se/omoss/index.html, accessed September 2007.

available in long-term planning and infrastructural actions. The association representatives are able to convey their assessments.

5.3.4.2 Resenärsforum[101] *(Passengers' Forum)* Resenärsforum sees itself as a lobby group for passengers in public transport. The organization works for the various forms of public transport – rail, bus and ferry services – in equal measure. At the annual meeting on 12 March 2005, the members decided upon a new orientation. Previously, the organization had been known as 'Modern Järnväg'. Resenärsforum is a member of the Swedish Konsumentråd, of the European Consumers' Union (BEUC) and the European Passengers' Federation (EPF). The organization works closely together on a national level with 30 local groups. There are c. 4,000 members. Members can be individuals, local groups or other bodies. In all issues relevant to passengers, the organization works closely with Banverket and Vägverket. Both offices are committed to being aware of all users' concerns.

5.3.4.3 Konsumentverket/Ombudsperson[102] As a national consumers' association, 'Konsumentverket' (KOV) represents the interests of consumers in various areas. This includes consumer information, product safety, and advertisement. The parliament and administration determine the parameters of consumer policy. Their work is federally supported. The association is directed by Håkan Sandersjö, who at the same time serves as Ombudsperson for consumers.[103] Konsumentverket has c. 180 employees. The organization deals with transport companies, for example the establishment of their general terms of carriage. The consumer protectors negotiate pertinent agreements with the companies. For this reason, there were talks in 2002 with SL and SJ about changes in their terms of carriage. There were further talks with the trade association SLTF[104] about a 'voluntary quality system'. In information from 2002, Konsumentverket described talks with SLTF as difficult. Particularly possible choices of compensation regulations lead to controversy.[105]

---

101 Website: http://www.resenarsforum.se, accessed September 2007.

102 Website: http://www.konsumentverket.se, accessed September 2007.

103 This unification of positions has existed since 1975. Konsumentverket's current duties and responsibilities were determined in statutory ordinance (1995:868) in the year 1995.

104 SLTF stands for 'Svenska Lokaltrafikföreningen'. The association is a conglomeration of regional and local transport operations. All 21 public transport authorities are members and 80 other transport companies. Eight employees coordinate SLTF's work.

105 http://www.konsumentverket.se/mallar/sv/artikel_datum.asp?lngArticleID=2312&lngCategoryID=1082, accessed September 2007.

5.3.5 Passengers' Rights and Customer Service

Passengers' and customers' rights have been an intense subject of discussion in Sweden since 2001. At that time, passengers' rights and participation possibilities were on the agenda.

On 17 May 2001, the government founded a commission called Public Transport with People at the Centre in order to analyse the current situation of public transport as a whole (kollektivtrafik). The commission had a total of 18 members and the following primary tasks:

- to analyse public transport from the customer's perspective;
- to analyse existing possibilities and problems;
- to foresee a vision for the time until 2020.

Involved as members in the commission were experts from the train industry, specialists in transport policy and representatives from Konsumentverket.[106] The final report was presented on 13 June 2003.[107] In approximately 500 pages, the commission presents its recommendations for public transport in the future.

Another commission produced an analysis especially for the rail sector. The government's corresponding resolution was made on 31 May 2001.[108] The starting points were the COTIF accord for international rail transport, the market re-opening for rail transport through the EU and an analysis of the various legal foundations in national railway law.

The final report for the rail sector was presented in December 2003.[109] In this report, the commission suggested that passengers should be paid compensation for losses. As an example, they cited theatre tickets, which can no longer be used if means of public transport are delayed. If the operator of the infrastructure was responsible for the delay, they would be liable for the losses.[110] Unified and nationwide information for all modes of public transport were to be offered. The legal foundations should be effective for all of public transport.[111]

---

106 List of participants at http://www.regeringen.se/content/1/c4/19/00/677cee8d.pdf, accessed September 2007. Konsumentverket was represented by Lotta Danin (until December 2002) and afterwards by Tove Källberg.

107 The final report is available under SOU 2003:67 on the Internet at http://www.sou.gov.se or http://www.regeringen.se/sou, accessed September 2007.

108 In Swedish. 'Utredning om järnvägssektorns organisation m.m.', 2001:48.

109 Available under SOU 2003:104 on the Internet at http://www.sou.gov.se or http://www.regeringen.se/sou, accessed September 2007.

110 For this purpose a contract should be negotiated, i.e. between the transport company and the infrastructure operator.

111 Close coordination with the Commission for Public Transport is foreseen for this.

5.3.5.1 Legal foundations The legal foundations for rail transport are the Järnvägslagen (2004:519), the Järnvägsförordningen (2004:526) and the Järnvägstrafiklagen (1985:192). The General Railway Statute became effective on 1 July 2004. A further statute determines stakeholders' responsibilities and requirements in regional transport.[112] The Lag om säkerhet vid tunnelbana och spårväg (1990:1157) applies for underground trains and street trams. There is no special statute for passengers' rights in Sweden.[113] Liability for trains is regulated by the Järnvägstrafiklagen. Liability is essentially limited to passengers' death or injury. Rail operators are also responsible for injuries incurred while boarding or disembarking from a train or during the journey, as long as the train is responsible for them. Liability for delay or cancellation of transport vehicles is not legally regulated. Legal claims in cases of delay or non-arrival of transport vehicles can thus not be derived from existing laws.

5.3.5.2 General terms of carriage The general terms of carriage (GTC) are called *resevillkor* or *allmänna resevillkor* in Swedish. Transport authorities and transport companies have their own general terms of carriage. The transport association's or SJ's *ex gratia* payment policy is defined in this document.

Swedish Railways (SJ) offers passengers a travel guarantee. Details are defined in Point 11 of their general terms of carriage. The passenger is supported according to overall travel time and delay. Thus, telephone costs are covered and, in cases of more than one hour delay, costs for meals. For occasions of important travel, expenses can be calculated for additional means of transport.[114] Consequential losses are not replaced. Other points of the travel guarantee refer to return of monthly tickets and cleanliness on trains. The passenger has three months within which to establish their claim against SJ.[115] The travel guarantee has been in effect since 15 January 2004. In the first half of 2004 there were 35,000 customer complaints. SJ paid SEK5.6 million (€607,000) in reimbursement. Approximately 90 per cent of these passengers received compensation.[116]

5.3.5.3 Regulation of ex gratia *payments* Most transport associations offer ex gratia payment as a service guarantee for their customers. In the Skåne region, one has a claim for reimbursement if transport is at least 20 minutes late. The transport association Skånetrafiken[117] reimburses passengers with a

112 Lagen om ansvar för viss kollektivtrafik (1997:734).

113 Stockholm Ministry of Transport.

114 SJ cites the example of trips to marriages or funerals.

115 For a comprehensive description of the provision see http://www.sj.se/sj/jsp/polopoly.jsp?d=3470&l=sv#, accessed September 2007.

116 Trafik i mälardalen, edition 4/2004, p. 7.

117 Skånetrafiken plans and organizes public transport in the regions Schonen/Öresund/Malmö.

free ticket for one day. Alternatively, the transport association covers costs for taxi travel or a trip in a private car for a maximum of SEK500 (€54). For this, the passenger must fill out a form and send their ticket to the transport association within 60 days. The transport association guarantees a processing time of 10 days.[118]

For Stockholm, Storstockholms Lokaltrafik offers a travel guarantee on a voluntary basis. In cases where a transport vehicle is delayed more than 20 minutes, the passenger can, as long as there is no existing alternative with another means of transport, ride in a taxi. The costs are covered up to SEK800 (€86.76).[119]

Västtrafik is responsible for the Gothenburg metropolitan area. This transport association has been offering a travel guarantee since 1 February 2004.[120] The transport association had collected experience with its two years of regulation limited to the city of Gothenburg. A transport vehicle must be at least 20 minutes late to warrant coverage of costs for a taxi or private automobile up to SEK300 (€33). The transport association assures a processing time of two weeks.[121]

The industry association SLTF states that all transport associations will gradually introduce a travel guarantee. Currently already approximately 95 per cent of trips are covered. The terms of the guarantees are to be improved.[122]

5.3.5.4 Public Claims Office[123]/alternative dispute resolution There has been a system of ombudspersons in Sweden since 1809. This system is meant to review how laws are carried out by public servants, magistrates or members of the military. There are ombudspersons for various political areas.[124] The function of the ombudsperson for consumers is currently administered by the leadership of the Konsumentverket. There is to date no ombudsperson for passengers. The Public Claims Office (Allmänna Reklamationsnämnden (ARN)) provides alternative dispute resolution. Their primary task is to provide nonpartisan reconciliation of disputes between consumers and companies. ARN is also responsible for the travel industry. Other fields include banks, residences, boats and insurance.

118 This time is named as a 'normal case'.

119 The guarantee is limited in some situations (extreme weather).

120 The guarantee is limited in some situations (extreme weather).

121 The travel guarantee is described extensively in a broschure at http://www.vasttrafik.se/ sv/Startsida/resegaranti/, accessed September 2007.

122 More information at http://www.sltf.se/newsletter/down.asp?theYear=2006&monthnr=11, accessed September 2007.

123 Website: http://www.arn.se, accessed September 2007. See also Section 2.5.

124 Other political areas are equal opportunities, ethnic discrimination, children and the handicapped.

The processing time is around six months. There are no fees. The minimum amount for a dispute in the travel industry is SEK1,000 (€108)). For this reason, many cases involving public transport are excluded from this process due to the fact that the dispute is too low in value. ARN conducts its process exclusively in writing. Legal advice is not given. ARN only accepts claims in Swedish. The Marknadsdomstolen (Trade Affairs Court) handles questions concerning competition, antitrust and advertisement. The Konsumentverket can initiate a process for issues of competition and advertisement before the Marknadsdomstolen.[125] The legal foundation is a statute that specifically addresses issues of advertisement, competition and product safety.[126]

5.3.6 Conclusion: The Tradition of Consumer Protection and Dialogue-Oriented Policy

Change has characterized the situation in Sweden in past years. All stakeholders have thus been moving closer to each other over time and looking for new ways to create intensive cooperation. Comprehensive consultation papers from specialized commissions are submitted and, in part, carried out. The COITF Accord, changed in 1999, and EU plans for changed passengers' rights are stimulating the national discussion.

Passengers' rights demonstrate a divided situation. There are only limited legal claims and there is no legal regulation. However, there are many mobility and service guarantees that are linked to restrictions.

Consumer protection has traditionally been accorded a high value in Sweden. With the Konsumentverket and the Claims Office, consumers have limited opportunities to provide their claims any degree of force. A special ombudsperson for transport could further improve the passengers' position. An arbitration board as an extension to the ARN would make sense for supporting passengers with claims of under €108.

Passengers' interests are represented by two associations. Järnvägsfrämjandet is more focused on trains and Resenärsforum is engaged in transport in general. There is no fixed and institutionalized participation for these associations in most cases. However, early changes are recognizable. Banverket and Vägverket involve the associations particularly when dealing with sensitive issues.

Thus, the situation is mixed and there is a need for action by all of those involved.

---

125 Website: http://www.marknadsdomstolen.se, accessed September 2007.

126 An extensive version can be found in the 2006 annual report at http://www. marknadsdomstolen.se/MDsArsredovisning/MDsArsredov2006.pdf, accessed September 2007.

5.4 The Netherlands

Holger Jansen

5.4.1 Political and Geographic Context

The Netherlands is a parliamentary monarchy. The approximately 16.3 million residents (as of May 2007) live in a land area of 41,500 km^2. The capital is Amsterdam, the seat of government is The Hague. The country is divided into 12 provinces. The smallest administrative units are the 443 municipalities. The Parliament consists of two Chambers: the Senate (First Chamber) and the Parliament (Second Chamber). The administration is divided into three levels: nation, provinces, and municipalities. The government in The Hague is responsible for national matters. The provinces administer to the areas of environment, land use, power supply, and social welfare. The municipalities are responsible for local matters.

5.4.2 Transport Market and the Organization of PT

The ministry responsible for national transport is called Ministerie van Verkeer en Waterstaat (Ministry for Transport, Water Management, and Public Work). This Ministry coordinates the transport providers for roadways, waterways, air travel, public transport, and bicycles. The office consists of seven General Directorates. The General Directorate for Passenger Transport is responsible for public transport.[127]

Administrative responsibility for public transport with regional importance lies with the 12 provinces and the seven metropolitan areas, which are called Regioplus.[128] The legal foundation for how the stakeholders work with one another is the statute for public passenger transport which came into effect in 2001.

As the central coordinator, the organization Kennisplatform Verkeer en Vervoer (KpVV)[129] supports local municipalities. The KpVV was founded as a consortium with the Centrum Vernieuwing Openbaar Vervoer on 1 January 2005. The organization has the following tasks:

- implementation of the goals set fourth in the Passenger Transport Law;
- development and preparation of information for tender offers;
- renovation of public transport;

127 Profile, goals and tasks of the Ministry are available at http://www.verkeerenwaterstaat.nl, accessed August 2007.

128 These regions connect the city and surrounding communities. More information can be found at http://www.stadsregios.nl, accessed August 2007.

129 Website: http://www.kpvv.nl, accessed August 2007.

- individual advising of transport authorities;
- workshops, yearly congresses, yearly reports;
- development and implementation of quality studies in public transport (OV-monitor and Klantenbarometer).[130]

KpVV's work deals with all modes of public transport. This particularly includes buses, street trams, underground trains, regionalized rail transport (delegated to the provinces) and public transport on waterways. The KpVV particularly supports transport authorities that invite tenders for their transport service.

The rail network measures c. 6,500 km. Since 2003, network and transport operations have been separate. The organization ProRail has been responsible for the maintenance and organization of the network since January 2003. Two thousand six hundred employees serve the rail network. ProRail is publicly owned. The market share of public transport[131] was 13 per cent in 2001, of which 8.7 per cent was trains. Other means of public transport had a market share of 4.2 per cent.[132]

The largest provider in rail transport is Netherlands Railways (Nederlandse Spoorwegen (NS)). In 2004, NS transported c. 1.1 million passengers per day. Domestically, NS transported a total of c. 320 million customers in 2004. The train companies Arriva and Syntus serve some regional train routes. A total of 13 companies provide transport services locally on rails and streets. In international transport NS cooperates closely with the trains in neighbouring countries and is a member of the common train network Thalys.

A new Passenger Transport Law came into effect on 1 January 2001 effect (Wet personenvervoer 2000 – Wp 2000).[133] The government was pursuing two main goals with this law: first, the level of cost coverage in public transport was to be higher. Second, public transport as a whole was to be made more attractive. The law allows for strong competition in the transport market. Two years after coming into effect, 35 per cent of transport services were to be tendered. For municipal transport operations, there was a deadline of 2006, which the Ministry extended to 2009. Originally, all public transport services were to be tendered by 2007 according to the law. In the tender invitations, the transport authorities define requirements for LPT transport companies. These include, among others:

130 A report on customer satisfaction in public transport can be found at http://www.verkeerenwaterstaat.nl/kennisplein/uploaded/AVV/2006-10/177376/KB-OV.PDF, accessed August 2007.

131 The reference figure is the passenger-km.

132 More information at http://www.rws-avv.nl/pls/portal30/docs/5455.PDF, accessed August 2007.

133 The law was passed in May and July 200 by the chambers of the Parliament.

- fares, determination of fares, timetables, issuances of invoices;
- accessibility for people with limited mobility;
- aspects of safety for personnel and passengers;
- participation possibilities for consumer associations;
- companies' information responsibility to transport authorities;
- bonus/malus regulation in transport contracts depending upon quality (for example, punctuality).

Every transport authority fundamentally only allocates concessions for their own area. In transport networks that extend beyond their own area, the transport authorities decide the form of the coordination. Thus, a transport authority can transfer responsibility for a determined line. It is also permitted to allocate two concessions to one company. Close coordination between the transport company and the transport authority is required. In some cases, the Ministry of Transport can intervene in a regulatory capacity. The transport authorities create their own fares. The Transport Minister is, however, authorized to create a national base fare that is effective for all means of transport, for example, with a smart card or strip ticket. Otherwise, the level of the fare is decided by the respective transport authorities.

5.4.3 Transport Planning and Participation

Participation possibilities for consumer and passenger associations are defined in the Passenger Transport Law. The law requires every transport company to process passenger complaints. The companies must inform passengers of their possibilities for issuing complaints. Every passenger has the right to know how his or her complaint is processed. Every company must have an association with an arbitration board. This board acts if a complaint cannot be satisfactorily settled between the passenger and the company.

Governmental offices and companies related to LPT are required to engage in regular dialogue with consumer organizations. This applies, for example, to timetables and fares. The consumer organizations are, according to law, treated as competent and constructive dialogue partners. Their recommendations must be addressed by governmental offices and transport services.

Consumer organizations, however, do not only play a role in current operations, but also in the procurement of tenders. LPT authorities must consult consumer organizations in the issuance of concessions and address their recommendations.[134] The law names the participation of associations and/or organizations in Article 27. The article states that the groups must have a relationship with the passengers or with public transport. This means legally that at least one meeting annually must take place between the organization and politicians, as well as another meeting with the transport company.

134 The details are defined in article 44 of the Passenger Transport Law.

For long-term planning, the Ministry of Transport creates a Nota Mobiliteit. All modes of transport are included in this plan. The current one is in effect until 2020. The provinces create additional plans. The administration and the Parliament passed the nationwide Nota Mobiliteit in 2006. The plan is publicly financed. It is presented in detail on the project website.[135] A short summary is also available in English.

5.4.3.1 Locov[136] Locov stands for Landelijk Overleg Consumentenbelangen Openbaar Vervoer. In English, Locov can be translated as 'National Support Centre for Consumer Issues in Public Transport'. This national organization is a discussion and consulting platform between politicians, transport providers and organizations for public transport issues with national relevance. The Ministry for Transport finances the organization, which was founded on 1 September 1998. In addition, the Ministry of Transport, the Nederlandse Spoorwegen (NS) and a total of six other consumer organizations are also represented. Among them are the Consumentenbond and ROVER.

5.4.3.2 Stichting OPC and OV-platforms[137] OPC stands for 'Ondersteuning collectief Personenvervoer aan Consumentenorganisaties'. The main goal of this organization is to regionally coordinate and implement the participation possibilities created for consumer associations by the Passenger Transport Law (see Section 4.6.4). Contact between the individual organizations is also to be maintained. OPC is a conglomeration of several transport and passenger organizations. This includes ROVER, organizations that deal with seniors, bicycles, and people with limited mobility. Their work is financed by the Ministry for Transport. For the regional and local levels in local public transport (LPT), there are advisory boards (OV-platforms), which must be included in issues of tender procurement and timetable changes. There is a legal claim for participation, as long as LPT has already been tendered. Several consumer and user groups work together regionally on the boards. The national OPC supports and coordinates the work of local platforms.[138]

5.4.4 Interest Representation and Lobby Groups

5.4.4.1 ROVER[139] ROVER is an abbreviation for 'Reizigers Openbaar Vervoer'. The translation is 'Travellers in Public Transport'. The organization

---

135 Website: http://www.vananaarbeter.nl/NotaMobiliteit/, accessed August 2007.

136 Detailed information at http://www.locov.nl, accessed August 2007.

137 Stichting OPC can best be translated as 'Foundation for Support of Public Transport and Consumer Organizations'.

138 More information about national OPV is available at http://www.stichting-opc.nl , http://www.rocov.nl and http://www.lijn7.info, accessed August 2007.

139 Website: http://www.rover.nl, accessed August 2007.

is an independent passengers' association. ROVER has c. 7,000 members in The Netherlands and was founded in 1971. The organization is an advocate for all public means of transport and works with all stakeholders in transport politics and with transport companies. ROVER is active in the whole country and has founded special work groups for specific issues. The association's magazine is called *De Reiziger* and is published six times annually. ROVER sees itself as an interest representation group for customers of public means of transport.[140] The association's main goal is to shorten travel times, make fares more attractive and create a more reasonable relationship between transport companies and customer complaints and suggestions. This is meant to make public transport a viable alternative to individual motor transport.

The association is becoming notably more strongly involved in planning and development processes. In some cases, a legal claim to participation exists according to the Passenger Transport Law.[141] ROVER advocates institutionalized participation in cases with international connections.[142] The passenger association accepts customers' questions and suggestions and discusses them with transport companies and politicians. The association awards an annual Reizigersprijs. This prize is meant to honour special performance in public transport. Anyone can win it: a company, a governmental office, a bus or street tram driver, a conductor or passenger who has improved public transport. There is a new theme every year. In 2006, the prize was won by a citizens' union, in 2007 by a community for its engagement in linking trains and bicycles.

5.4.4.2 Consumentenbond[143] The Consumentenbond sees itself as an interest representation group for passengers and of all means of transport. The organizations aims to provide a critical voice to the government's transport policies. The Consumentenbond understands itself as representing the general interests of all consumers. The organization was founded in 1953 and has 620,000 members. It belongs to the umbrella group Consumers International (CI) and the Bureau Europeen des Union de Consommateurs (BEUC, Brussels). The Consumentenbond is a member of Locov.

140 Three million passengers use public transport in The Netherlands daily.

141 Participation possibilities exist for national transport. International transport is excluded from this. A comprehensive evaluation from the perspective of ROVER is available at http://www.rover.nl/downloads/index.php, accessed August 2007.

142 http://europa.eu.int/comm/transport/rail/passenger/doc/rover1-nl.pdf, accessed May 2007.

143 Website: http://www.consumentenbond.nl, accessed August 2007.

5.4.5 Passengers' Rights and Customer Service

5.4.5.1 Legal foundations The legal foundation for transport companies' liability is the Civic Code of Law (Burgerlijk Wetboek). According to Article 8:108, liability for delay is excluded. The General Terms of Carriage (Algemene voorwaarden) in transport refer to the determinations in the Civic Code of Law. Further liabilities, such as loss of baggage or injury of travellers, are not dealt with in detail here.

5.4.5.2 General terms of carriage Every transport company determines further regulations for the legal relationship between passengers and companies in its terms of carriage. In the terms of carriage set fourth by Nederlandse Spoorwegen (NS) Article 13.3, liability for delay is excluded with reference to the existing legal ordinances. NS refers, however, to *ex gratia* regulations 'money back in cases of delay'. In Article 15 the addresses for passenger complaints are named. The customer must contact NS within four weeks. The passenger is informed in Article 15.2 that they can contact the Geschillencommissie Openbaar Vervoer[144] if they are not satisfied with the answer from the NS. In the terms of carriage of other public means of transport, liability in cases of delay or cancellation is excluded with reference to the Civic Code of Law.

5.4.5.3 Ex Gratia Regulations Nederlandse Spoorwegen (NS) reimburses half of the fare in cases of delay longer than 30 minutes. In cases of more than 60-minute delay, the full fare is returned. There is no legal claim according to law. Time limits or reimbursement figures are not defined in the general terms of carriage (GTC). Regulation dates back to an agreement between the Ministry of Transport and NS and came into effect on 15 February 2001. Passengers' legal claims are limited by situations beyond the transport company's control. NS has only invoked this exception in a few individual cases. On one day there was a heavy storm with numerous disturbances in the network. On another day, employees were striking, which the media informed passengers of beforehand.[145]

The compensation is transferred to the customer. In 2002, there were c. 800,000 grievances regarding the 'money back guarantee in cases of delay', in 2003, c. 600,000.[146] There are additional regulations for long-term tickets, bicycles, transfer tickets, tickets for specific long-distance trains and tickets for

144 Website: http://www.sgc.nl, accessed August 2007.

145 Inquiry with NS.

146 Other grievances with NS in the 2002: c. 28,000; 2003: c. 17,000. These include damages due to delay that are not included in the guarantee. *Ex gratia* payment is made by NS in individual cases of this kind (for example, coverage of taxi costs).

dogs. In such cases, proportionate amounts are reimbursed.[147] The minimum amount for reimbursement is €2.20. Further compensation regulations are at the discretion of the respective companies.

Some companies offer regulations on a voluntary basis. Thus, for example, the Rotterdam transport operation RET promises a guaranteed connection, as long as it is the first or last possible connection.[148]

5.4.5.4 Stichting Geschillencommissies voor Consumentenzaken (SGC)[149] A special organization, the Foundation for Arbitration in Consumer Issues (SGC, see Section 2.5.5) helps passengers in cases of dispute arbitration with transport operations.[150] In 30 different commissions, SGC experts deal with the most various topics. Telecommunication, automobiles, banks, and legal councils are among these topics, as well as public transport. The goal is to arbitrate disputes between consumers and providers extra judicially. In 2006, 12,990 consumers used SGC:[151] 186 grievances addressed public transport. Approximately 30 full-time employees work for the organization.

There is a commission for every topic addressed by the arbitration board. The chairman is a legal practitioner. Each has a representative from the consumers' associations and from the provider. The SGC only deals with problems within The Netherlands. The public transport commission applies the following policy: the customer must first issue a grievance in writing to the respective transport company. An answer must have been due for more than four weeks. The transport company must be a member of the association Mobis.[152] The SGC has been dealing with passenger grievances since 1 January 1997. The foundation of their work was an agreement between SGC, the Consumentbond, NS and Mobis.

In specific cases, the commission is not active. This includes cases where the value of the grievance exceeds €3,403.35. Further limitations are stated in respective regulations. SGC presents model cases on its website.

---

147 Detailed information at http://www.ns.nl under 'Geld terug bij vertraging' and http://www.schlichtungsstelle-nahverkehr.de, accessed August 2007. The arbitration board for local transport inquired with NS through the Consumers' Centre NRW about previous experiences with compensation regulation.

148 More at http://www.ret.nl/?id=93, accessed August 2007.

149 Website: http://www.sgc.nl, accessed August 2007.

150 SGC's work deals equally with passengers of NS and other public means of transport.

151 SGC Annual Report, p. 24. Available at http://www.sgc.nl/Documenten/afbeeldingen/ Jaarverslag per cent202006 per cent20def.pdf, accessed September 2007.

152 Mobis is an association of transport companies in public transport. Along with supporting SGC's work, Mobis also has its own platform for questions of safety in public transport. Members of numerous organizations and politicians work as part of this platform. More information at http://www.knv.nl, accessed September 2007.

5.4.6 Conclusion: Developed Participation and Good Examples for Customer Service

The participation possibilities for consumers' organizations in The Netherlands are good. There is an array of legal claims that are legally anchored to passenger transport. Other important factors are the organizations and platforms that support smaller organizations with more limited means. This addresses especially the commonly encountered imbalance in interest representation groups between the service provider's and the customer's side. While service providers' interest representation groups are well-organized, this is not always the case on the passengers' side. Financial engagement from the Ministry for Transport would thus be a welcome addition.

There is a mixed situation for passengers' rights. Legal claims are, at first glance, limited. The NS's regulation is anchored in the general terms of carriage – it is, however, exemplary. In cases of delay of more than 60 minutes, NS reimburses the full fare. In the same situation, other train companies would reimburse – if anything – only part of the fare. There is still, indeed, need for action for the remaining modes of PT. Some companies promise specific standards. There is still no unified policy. Thus, regulations in tenders offer good opportunities for unified standards in particular regions.

The extra-legal settlement of disputes through an arbitration board should certainly be named as a positive factor as well. Transport companies must refer to such organizations. This too is a large step forward in comparison to many other countries. In cases of customer grievances, companies must address these in accordance with the law. The Netherlands provide many kinds of participation possibilities for passengers that are transparent and comprehensible.

5.5 France

Uwe Böhme[153]

5.5.1 Political and Geographical Context

France, like other European countries, is divided into many administrative levels whose basic structure has scarcely changed since the Revolution. Until 1982/83 governance in France was still strongly central. With François Mitterand's entry into government and the final decentralization, the country's authority was divided into three further levels:

1. 22 regions, governed by a Conseil regional;

153 I would like to thank Patrick André, Louis Kaeppelin and Martin Schiefelbusch for the material and commentary they provided for this text.

2. 100 Départements (Departments) – comparable to the British County Councils; subdivided into Arrondissements (2006: 329) and Cantons for administrative purposes;
3. Municipalities (communes, 36,571) governed by district councils, which also choose the mayor.

This ordering does not, however, reflect a hierarchical system because the levels do not have authority over one another and there is a principle of self-administration. In terms of public transport, responsibility is divided as follows:[154]

- National:
  - general legislation;
  - authority over national transport services;
  - determination of the scope of transport policy (Masterplan);
  - financial support of transport authorities;
  - definition of safety standards and checks;
  - collection of statistics and performance of studies;
  - determination of limits for ticket fares;
- Regional:
  - organization of regional rail transport;
  - creation of regional development plans (Masterplan);
  - allocation of contracts to transport companies;
- Departmental:
  - organization of school transport;
  - production of Masterplans on the Departmental level;
- Municipalities and/or federations thereof:
  - organization of public transport where a special transport sector has been defined;
  - production of local transport plans.

Because of the transference of decision-making authority to multiple levels, the state, as the highest administrative level, must see to it that national interests are addressed. Finally, the sub-national levels determine over 75 per cent of the total budget for public investments.[155] In addition, the Office of the Prefect was instituted to represent the interests of the state in the Departments and regions. The prefect of a region is usually also the prefect of the most important Departments in the region. Thus even after decentralization, the state still plays an important role in terms of administrative and economic interests.

154 See Direction des Transports Terrestres (DTT) (2003), p. 108.
155 See DTT (2003), p. 104.

5.5.2 The Transport Market and the Organization of PT

5.5.2.1 Transport authorities Since 1982, when the 'loi d'orientation des transport intérieurs' (Domestic Traffic Law (LOTI))[156] came into effect, special transport organizing authorities (*autorité organisatrice* (AO)) have been responsible for the overall planning and organization of public transport within defined areas.[157] Their tasks include, for example, the creation of transport development plans, the determination of transport offerings and their supervision, as well as the financing of infrastructure. In most cases (89 percent), they have, however, transferred the tasks to private companies. There are currently over 250 such 'autorités organisatrices des transports urbains' (AOTU) in France, which makes the status of an AO quite variable – the simplest organizational entity is a federation of municipalities that form a cooperational partnership.[158] GART[159] was created as a national umbrella group. It has 252 members and offers a space for mutual exchange of experiences.

5.5.2.2 Rail transport The state transport company SNCF[160] serves almost all rail transport including the high-speed train TGV,[161] regional trains as well as trans-European rail transport (Eurostar, Thalys). Long-distance bus transport is also part of SNCF's service, but is however operated by daughter companies. The rail network is owned by the infrastructure operator Réseau Ferré de France (RFF), which is independent of the SNCF.

In terms of the service offering, there are, however, major differences between the Paris metropolitan area, with its very tight network and the rest of the country, where there is only limited regional transport. Moreover, in the building of the rail network, France is primarily concentrated on the high-speed network of the TGV and thus neglects other areas such as transport of goods. The truncation of the goods transport railways in the modal split is often categorized as the most difficult problem in transport politics.[162]

5.5.2.3 Local public transport and regional bus transport The world of local public transport is dominated by three private companies, which share more

156 Only outside Paris; inside Paris, the role of AOTU is given through special assignment to STIF, see Section 5.5.2.3.

157 PTU = Périmètre de transports urbains.

158 EPCI = Établissement public de coopération intercommunale.

159 GART = Groupement des autorités responsables de transport.

160 SNCF = Société Nationale des Chemins de fer Français.

161 TGV = Train à Grande Vitesse.

162 See Potthast and Dienel (2003), p. 10.

than 74 per cent[163] of the market. These include Kéolis, Transdev and Connex/ Veolia Transport. In smaller cities, other private companies are also active. These companies have amalgamated into the association AGIR (Association pour la Gestion Independante des Reseaux), which also provides support in tender award procedures. In total, approximately 90 per cent of transport services are provided by private companies.[164] Publicly owned companies only exist in the cities of Paris and Marseille.[165] Organizational and plenary responsibility is held by the TA. These responsibilities are determined by the transport contracts, including the frequency of service, the pricing[166] and the vehicles to be employed.

Because PT deals with subsidized transport (income from fares covers c. 30 per cent of operational costs),[167] the AO also possesses diverse control and sanctioning mechanisms, which address criteria such as cleanliness and punctuality.

School transport dominates transregional bus transport. For this reason, in times of lower demand, smaller buses or taxis are used. This mode of transport is organized through the Departments and is primarily carried out by private companies.

As a whole, the last 15 years have seen a significant improvement for transport services with the application of new technologies and increased investment in the sector. For example, several new street tramlines have been opened and several train stations have been expanded into intermodal stations (for example, in Nantes).

The PT Tax (Versement Transport), which was first instituted in Paris in 1971, has had a significant influence on capital investment. Under specific conditions, a TA can collect this tax from private and public institutions with more than nine employees. Its implementation is justified by the fact that employers profit from public transport services, i.e. through better accessibility. This tax has been instituted in most PTU and now comprises the most important financial resource, financing c. 40 per cent of operational and investment costs for LPT in France.[168] Because the tax is oriented toward economic strength and not on the costs of investment projects, particularly affluent communities tend to invest in overdesigned projects.[169]

Because of Paris' special situation (11.1 million residents and 1,281 municipalities, at the same time a Department and district) there is a TA there

163 Seventy four per cent of the network with 80 per cent of vehicle mileage, see CERTU et al. (2005).

164 CERTU et al. (2005).

165 See Berschin (2000), p. 101.

166 Within the restrictions set by the state.

167 See DTT (2003), p. 37.

168 See Klein (1996), p. 57.

169 See Rogge (1999), S. 55; Klein (1996).

with special functions whose 34 members are representatives of the State, the region, and the Department. The range of responsibility covered by the Syndicat des Transports des Ile-de-France (STIF) comprises the organization of transport in the Paris metropolitan area. This includes coordination with the state train operator SNCF, the transport company RATP,[170] which operates the Paris Metro as well as over 90 private train companies in surrounding area. Furthermore, the STIF coordinates the allotment of the PT Tax.

5.5.3 Transport Planning and Participation

While there is no specific designation for formal and informal participation in France,[171] there are developments toward creating a legal foundation for coordination that includes those affected. As in Germany, there are various planning instruments, determined by the various administrative levels, which address the interests of the country, region, or district, as well as those of the State and of transport planning. For cities with more than 100,000 residents, a transport development plan (*plan de déplacements urbains*) is required,[172] and a quota of stakeholders must be included in the creation of and changes to this plan including consumer associations, the TA responsible for the area and environmental protection agencies. In the creation of a *plan for urban development* for a municipal area (*plan local d'urbanisme*) as well as plans for land use (*plan d'occupation des sols*), citizens are included in various ways.[173]

Another possibility for participation is designed to give citizens a part in the development of especially larger infrastructure projects (for example, Port 2000 in Le Havre or highway A32 between Metz and Nancy). The foundation of this was set through the Bianco circular and the resulting foundation of the Commission Nationale du Débat Public (CNDP) for the implementation of participatory practices.[174] The goal is to achieve early recognition of approaching problems that will affect the project and thereby reduce the risk of conflict between the constructor and the public. When substantial effects on the environment or on land use can be expected, the citizens have the possibility during the entire practice to input their own concerns and ideas. Since 2002, the range of responsibility has expanded and the Commission now has the task of ensuring adequate citizen participation and assuring that all participating stakeholders take part.[175]

170 RATP = Régie autonome des transports Parisiens.

171 See Töllner (2002), S. 110.

172 Müller (2004), p. 375.

173 See Töllner (2002), pp. 85 et seq.

174 As an overview of the development of the *Débat public*, see http://http://www.debatpublic-francilienne.org/debat/histoire.html, accessed June 2007.

175 Ssee Rui (2004), p. 131.

As a whole, citizens' use of participation possibilities has often remained on a low level. The reasons for this are various and determined particularly by history. Until the beginning of the 1980s, France was a centralized state with enormous power. At that time, the public demanded solutions to social problems from the state.[176] Even after decentralization, citizens were hardly perceived as participants, and cooperation rarely took place because people had the opinion that a citizen is represented by his officials.[177] The understanding of roles has indeed been fundamentally changing for several years; however, this could be a reason why, compared to Germany, citizens' initiatives in France are rarer and terms such as public work and citizen participation are not accorded the same value.[178] At the same time, parts of the public engage in interest groups for the improvement of PT, which will be presented in brief in the next section.

There are a few other participation possibilities available to these associations, in that they are represented in national, regional and Departmental *transport committees*.[179] These bodies, however, do not address public transport in particular, but are mixed committees for transport issues of all kinds, which include representatives of parliaments, administrations, transport authorities, transport companies, trade unions, chambers of commerce, and consumer associations. The national committee has 193 members.

The committees deal with, first of all, issues of transport policy that can affect political affairs or their own interests. They thus serve to provide intercommunication and consultation for policy makers, without having any decision-making power of their own. The national and Departmental boards, however, also have a supervisory function in that the committees advise about transport companies' offences against trade legislative provisions and decide the type and extent of sanctions to be imposed. The national committee thus serves as an appellate court for decisions on the Departmental level.

In the course of SPNV reform in France, the so-called *route committees* (Comités de lignes) have been used as an additional forum since the late 1990s.[180] The foundation for this is the Law for Local Democracy from 2002,[181] which, however, does not require authorities or operators to form such a committee. They are to provide small-scale dialogues (if applicable, about their own train routes) between passengers and providers, and to discuss service development quality of the respective route(s). Members are representatives of communes,

176 See Rogge (1999), p. 52.

177 See Töllner (2002), p. 165.

178 See Rogge (1999), p. 52.

179 Conseil national des transports, comités régionaux et departementaux des transports.

180 See Schnieders (2004).

181 Loi n° 2002/276 du 27 février 2002 relative à la démocratie de proximité.

transport companies, regional train personnel and consumer associations – also, in part, individuals and passenger representatives.

5.5.4 Interest Representation and Lobby Groups

Fundamentally, a law from 1901 allows citizens to form unions and lobby groups. On this foundation, citizens have joined together, mostly on the level of larger cities, with common concerns about their environment in a broader sense, or about the travel sector. Some interest groups have coalesced into larger organizations in order to expand their influence and operate on a national level. Most finance the work of their associations with membership dues and donations, also partly through state subsidies. Under specific conditions, associations receive a licence that gives them the right to advocate consumer concerns in court. On the national level, 18 such licences were given in the area of consumer representation.

Probably the most influential organization in the transport sector is the umbrella group of passenger associations FNAUT (Fédération Nationale des Associations d'Usagers des Transports), which has an aforementioned licence. In the association, founded in 1978, railways were at first the best-represented group, the federation now includes c. 150 associations with a total of 40,000 members. Their main activities are concentrated on rail transport/long-distance transport whereas concerns of bus and ferry transport are largely not addressed. The focus is placed especially on representing passengers in cases of complaints of dissatisfaction with transport companies SNCF and RATP (see Section 2.5), on passenger information, and on the implementation of general passenger interests. Because FNAUT has a seat on the RATP board of directors, the basic conditions for good communication exist. Indeed, an exchange of information does take place, but implementation is often sluggish.

5.5.5 Passenger Rights and Consumer Protection

5.5.5.1 Quality certification for transport services In all sectors of transport, there are companies that offer certification of transport services on a voluntary basis. In 2001, 50 of the 87 member companies of the umbrella group UTP (Union des transports publics) set criteria for quality in their transport contracts which were determined in cooperation between the respective transport companies and consumer associations.

A transport company receives a certificate through the Approval Association for Certification (AFNOR)[182] if the activities they have introduced fulfil the criteria. A subsequent check of the service quality is meant to ensure

182 Association française de normalisation (French Association for Standardization).

that service corresponds to the certificate's criteria. Otherwise, the transport company risks having its status as a quality-oriented company revoked. Because the certificate signifies a high level of quality to the passenger that he can rely upon in the long-term, it holds a special value and an incentive for transport companies to fulfil the stipulated quality criteria. Since the first certificates in 1993/94 were inaugurated in Paris and Lyon, not only service quality but also cooperation between the various stakeholders has improved. The inclusion of passengers has seen to it above all that passengers profit from positive developments in PT. Since this time the practice of certificates has also been based on legal regulations (cf. Section 3.5.1).

5.5.5.2 Customer charters and service guarantees Since the principle of the quality certificate has been established, several customer charters have been instituted starting in the year 2000. After long talks with consumer organizations, the French state train SNCF also released a customer charter for national rail transport (TGV and Corail). According to this charter, in cases of more than 30-minute delays, passengers have a claim to a travel coupon worth a third of the ticket if the SNCF was responsible for the delay. In contrast to quality certificates, there is no legal basis for these regulations.

Furthermore, there are several exclusion criteria such as strikes, which are not at all rare in public transport. In order to ensure service in these cases, since July 2006 transport companies have committed themselves to establish a replacement service and to extensively inform their customers in a timely manner (48 hours before the strike).[183] Such a *Service garanti* thus determines to what degree replacement service is to be provided and what penalties could be charged if a company does not meet its requirements. President Sarkozy announced that in the future companies will also be legally required to provide a replacement service in the case of a strike.

5.5.5.3 Complaint management and alternative resolution of disputes Passengers in France have to date not yet had a legal claim in cases of cancellation and delay and only a few voluntary guarantees. One exception is city bus transport in Cherbourg where the bus company Zéphir Bus started a push for quality in 1995. Starting with only five minutes of delay, the passenger receives financial compensation if the company is at fault.

Many transport companies have internal standards of how customer complaints are to be handled. Most refer to time limits for addressing complaints. Furthermore, customers of RATP and SNCF have the possibility of appealing to one of the 18 nationally recognized consumer organizations (for example, FNAUT) in the case that they are not satisfied with the answer. These organizations are legally allowed to contact a neutral arbiter (*médiateur*) and resolve the issue on behalf of the passenger with the transport company

183 Similar arrangements also exist in Italy, see Section 5.6).

(see Section 2.5). In this way, it has been possible to extra judicially resolve over 80 per cent of cases that have to date been presented to the *médiateur*.[184]

5.5.6 Conclusion: Formal Modes of Participation

Through decentralization and the resulting transference of responsibility for organising public transport to special transport authorities, these bodies have gained the possibility to make decisions about improving PT directly on the local level. However, due to the complex administrative structure and its according difficulties in the determination of responsibilities, it has been difficult to develop ubiquitous and unified transport services. Especially in cases where routes do not cover the borders of a district, there are considerable shortcomings in terms of unified cycles and fares as well as good transfer options.[185]

Accordingly, various developments during the last decade have sought to improve the basic conditions in public transport despite mass motorization and population growth in metropolitan areas. The PT Tax in particular, which has provided important financial resources, has largely contributed to service improvements. Furthermore, the passenger has profited through the introduction of service guarantees, at least in rail transport, because he can receive financial compensation from the SNCF in cases of delay. The award of quality certificates has provided an incentive for transport companies to keep their services at a high level in the long term. Because passenger representatives participate in the creation of criteria, their interests are addressed as well.

Indeed, all too often it happens that despite many participation possibilities, participation is a consequence of highly engaged local interest groups, because still too little awareness exists for consistent citizen participation in the community. With the choice of a citizen representative, decision-making power is transferred; thus political decisions are the will of the people. Barring spontaneous protests against infrastructure projects or participation in passenger associations, it is hardly possible to influence decisions.[186]

However, this social tradition has been fundamentally changing recently, and one can recognize in France a trend toward more citizen participation that comes close to a revolution, comparable to the fall of the Berlin Wall.[187] Thus participation seems the best-suited way to ensure democracy. The Bianco circular of 1992 gave important indications in this direction; whether this was actually the beginning of a fundamental change in mentality remains to be seen.

184 EUSG Consortium (2006), p. 95.
185 See Faivre d'Arcier (2003), p. 4.
186 See Varnier et al. (2001), p. 2.
187 See Rui (2004), p. 130.

5.6 Italy

Martin Schiefelbusch[188]

5.6.1 Geography, Politics and Transport

The Repubblica Italiana is one of the founding countries of the European Union and at the time of writing has 57 million residents. Italy re-emerged in 1861 after long having been divided into several previously independent states. Contrasts in cultural and social conditions as well as differing standards of living in the individual parts of the country can still be felt today. Large differences exist especially between the strongly industrialized North with its dense population centres of Milan, Turin and Genoa, and the rather rural South (called the Mezzogiorno).[189]

Along with the central state there are 20 regions and 103 provinces. For a long time, the national level was determinative in almost all fields of policy. In the course of various reforms[190] since the 1990s, the regions in particular have gained additional responsibilities and competences that stretch into the transport sector (see Section 5.6.3). The situation is thus significantly more heterogeneous now than just a few years ago.

There are, however, a range of features that, compared to the other countries described in this chapter, can be taken as characteristic for Italy as a whole. Among *spatial features*, these include the high proportion of mountains, which makes the transport coverage difficult, and the compact structure of many cities. *Political characteristics* include the instability of the (central) government and a political style that is more oriented towards personalities than programs in a strongly split party system. This often leads decision makers to take positions that are oriented toward confrontation and public efficacy, which are a stark contrast to an implementation of decisions, once they are made, that is often pragmatic and sluggish to the point of non-existence.[191] In the transport sector, this can be seen in an orientation toward large projects (such as highways or underground trains), which consistently

---

188 Along with the cited sources, this text is based on a series of expert interviews. For comments and encouragement, I should also like to thank Orietta Bittoni, Massimo Ferrari, Fulvio Pellicani, Trenitalia Direzione Generale Operativa Passeggeri and the Ufficio Servizio Ferroviario Regionale – Regione Lombardia.

189 As an overview see Pohl (1995).

190 See, for instance, Grasse (2004), Bramezza (1998), Dente (1999).

191 See Potz (1998), Grasse (2004), Drüke (2004), p. 19 as well as Vassalli (1998). Such models can be perceived in urban development as well, see Balducci and Calvaresi (2000), pp. 76 et seq. Sabelberg (1984), p. 37.

are delayed by frequent change in administrative and political leadership, complicated administrative procedures, and insufficient financial means.[192]

With 591 passenger cars per 1,000 residents, Italy has the second-highest *level of motorization* in Europe, and including the 11 million motorized cycles it takes the prime position. In its relationship to economic capacity (GDP per head), the role of MIT (motorized individual transport) seems higher than in comparable countries.[193] But the orientation toward automobiles that earlier shaped transport politics has differentiated, which has been influenced by ecological problems as well as institutional reform. International recognition of this view was made possible by the *shut-down of the inner city* for MIT in Bologna in the late 1980s.[194] Since that time, many Italian cities have introduced such traffic-free zones. However, there are more or less substantial exceptions and limited supervision which demonstrate the pragmatic understanding of politics.[195]

5.6.2 Features of Public Transport Services

The largest provider of PT services is the state rail transport company Trenitalia, which began in 2001 in the course of a train reform on the European level. The operator and owner of the railway infrastructure is Rete Ferroviaria Italiana (RFI); the previously integrated state-owned enterprise Ferrovie dello Stato (FS) now acts as a general holding company. Trenitalia itself is separated into divisions for long-distance passenger transport, regional transport, and goods transport. An additional 23 rail transport companies are active in regional passenger transport, mostly on their own lines.

Urban PT is generally offered by municipal companies which also operate other services, including parking space management. Underground trains and street trams play a comparably small role as the existing networks mostly have a limited range and/or qualitative deficiencies. Service standards, productivity of companies, and passenger satisfaction are generally lower in the South of the country than in the North.[196]

An essential feature of Italian PT is the pricing, which counts among the least expensive in Europe. In the 1960s and 1970s, fares were not raised in order to win more passengers and to make mobility possible for less-advantaged social groups. Even today prices in urban transport are 80 cents to €1 per

192 See Legambiente (2006), Osservatorio Territoriale Infrastrutture Nordovest (2007), detailed information on state rail transport: ISFORT (2006).

193 Figures from European Commission (2004), Tables 1.1 and 3.6.1–3.6.5.

194 See Apel (1992) and Pinardi (1994).

195 ISFORT (2005), pp. 7, 9.

196 See ACI and Eurispes (2006), Carminucci et al. (2005), ISFORT (2006b).

single ride, significantly lower than middle and northern European levels.[197] The subsidization of transport services in all areas is part of Italy's political tradition because it was often not possible even in the nineteenth century to build railways without partial public funding, due to limited industrialization.[198]

Despite low fares, numerous surveys show a high degree of dissatisfaction with public transport among Italians. Particularly criticized are unreliability, lacking information, cleanliness and coordination of different modes of transport with one another.[199] The train tends to receive somewhat lower ratings than other transport services. Public transport in Italy, even in comparison to other industries and other countries, is at the bottom of the popularity scale.[200] Alongside the aforementioned factors, this also demonstrates public transport's difficult starting point.

5.6.3 The Public Transport Reform since 1997

In the course of decentralization of governmental responsibilities, the 1997 Law 422/97 (also known as the Bassanini Law) transferred the planning and financing of local and regional transport to the regions, while simply local transport remains the responsibility of the provinces and municipalities.

The law implements, among other things, a European effort to liberalize the LPT market and draws some parallels to the rail reform and to regionalization in Germany and other countries. The fundaments of the law are:

- transference of the funding which was previously provided by the central government to the regions according to their proportion of transport service provision;
- sovereignty of planning: the possibility of procuring services through the regions that exceed this minimum offering;
- sovereignty of finances: determination of fares and decisions about provision of further funding to be done by the regions;
- separation between the levels of the purchaser (region) and the provider (transport company);
- transformation of transport companies into formally private businesses (plcs);

197 See BASIC-Consortium (2005), pp. 52–68, CNEL (2005), Carminucci et al. (2005), pp. 61 et seq.

198 See Maggi (2005).

199 EUSG-Consortium (2006a), p. 192, Carminucci et al. (2005), pp. 30 et seq., ISFORT (2006b), p. 36 et seq., ACI/Eurispes (2006), pp. 7–9.

200 In the last Eurobarometer study carried out on this topic, 41 per cent of Italians asked were dissatisfied with state LPT and 44 per cent with rail transport, in both cases the highest rate of dissatisfaction in the 25 EU countries. See European Comission (2004a), p. 16.

- public tendering of transport services.

Regions are required to pass a regional LPT Law that regulates the allotment of competencies between region, province, and municipality, practice and content of local transport plans, institution of inspecting authorities, and public participation (see below). Some of the laws also include propositions for further transport services such as non-scheduled transport, airport shuttles, and ferries.

Local authorities have thus taken over substantial responsibilities from companies and from the central government. In reality, there are however still cutbacks to be made in the implementation of the reforms to date. Thus passing the regional PT laws stretched into the year 2002, and the composition of new institutions is still not concluded.[201]

5.6.4 The Carta dei Servizi

The Service Charter (Carta dei Servizi), which only exists in this form in Italy, is an instrument for the formation of connections between transport companies and their passengers. It is a document that states the company's activities, standards and goals for quality, information for usage of the service, and possibilities for making complaints and receiving compensation.

The concept of a service charter was introduced in 1994 with a government initiative to improve the quality, transparency, and customer-orientation of public services – not just public transport. Every company is required to publish a service charter and to update it annually. The government developed detailed requirements for the content of the charter with its principle decision and the following sector-specific ordinances. It is incumbent upon the company to formulate regulations and determine standards that fit their respective situations. Table 5.1 shows a simplified version of the provisions for public transport.

Even information about the complaint and compensation possibilities to be offered by the company are found in the service charter. A positive feature in this respect are the 'low-threshold' services, such as the several information offices, displayed documents, and numerous free customer telephone lines which provide the opportunity to receive guidance and place criticisms (see Section 2.3).[202]

In its appendix, the ordinance includes a total of 17 tables addressing modes of transport which make detailed suggestions for indicators and survey practices for the quality criteria described in part 2 of Table 5.1. Table 5.2 shows an excerpt of such a description for national rail transport.

201 Detailed information on this topic is to be found in ISFORT (2006a) esp. pp. 18 – 40, CNEL (2005).

202 See Giammetti (2000).

Table 5.1 Structure of the service charter for public transport according to decision of the cabinet

Part 1: Basic principles	
Openness	Services provided without discrimination; accessibility and equal treatment for all users
Continuity	Duty to operate, required provision of information and replacement in cases of services disruption
Participation	Guarantee of dialogue with consumer associations
Cost-effectiveness	efforts toward efficient and effective service provision
Freedom of choice	Observance of the basic right to mobility
Part 2: Areas of service quality	
Technical safety	Travellers' personal safety
Reliability/punctuality	Cleanliness/hygiene
Comfort while travelling	Additional on-board services
Service for handicapped people	Customer information
Conduct of the personnel	Standards of the sales system
Integration with other modes of transport	Environmental friendliness
Part 3: Customer service and communication	
Personnel	Requirement to wear identification badge, neat appearance, and customer-oriented mode of communication
Customer service	Possibilities for making complaints, criteria for grievances, policy in case of personal injury or property damage
Quality control	Satisfaction surveys and practices for supervising determined quality indicators

Source: Schema generale di riferimento per la predisposizione della carta dei servizi pubblici del settore trasporti (Carta della mobilità) – Decreto 30.12.98.

According to the service charter's overall goal, customer participation in the design of customer service is to be addressed in the charter. Consulting with consumer associations about charter proposals is thus one of the formal participation possibilities and is regularly practiced. The governmental ordinance even mentions a general dialogue requirement. In any case, however, the participation is purely consultative, which the consumer associations see as insufficient.[203]

The service charter has also been looked at sceptically by user consumer associations and customers themselves. Their own experience of quality

203 Assoutenti Lombardia et al. (2005), p. 5.

Table 5.2 Examples of quality indicators in the service charter for rail transport

Area	Indicators
Technical safety	Deaths/injuries per passenger-km Judgement of safety in satisfaction surveys
Travel comfort	Average age of vehicle fleet Efficiency (person-km/space-km) Spatial conditions (m^2 per traveller) Cars with climate control (per cent) Judgement of comfort in satisfaction surveys
Integration	Number of train stations in airports/harbours, with P+R spaces Transfer possibilities to LPT Judgement of integration in satisfaction surveys

Source: see Table 5.1.

deficiencies and the disinterest of companies leads to a general mistrust of such declarations of intent. The mode of performance description (see Table 5.2) is also seen as rather unhelpful: the quality indicators, which are often determined on the company level, do not correspond with users' experience and thus seem non-reproducible to the point of potential manipulation. This means that additional 'trust-building steps' need to be taken by the service provider in order to allow the service charters more acceptance.[204]

Several Italian studies also come to the conclusion that the service charters have not yet fulfilled the expectations allocated them by policy. Among other things, the following aspects have been criticized:

- the use of technical and/or unclear language;
- the frequent lack of updates or comparative data from earlier periods;
- lack of public relations on the company's part;
- inconsistent, non-reproducible quality criteria;
- lack of sanctions for non-compliance with the standards defined in the charter;
- presentation of satisfaction levels without naming their methodological foundation.[205]

Legislation for service charters stipulates the requirement to send exemplar copies to a committee supervising the process. But one contributing factor to

204 See D'Alessandro (2000).

205 See Istituto per la promozione industriale (2003), Centro Studi Federtrasporto (2001), pp. 109 et seq., Prosperetti et al. (2000), pp. 197 et. seq., Assoutenti Lombardia et al.. (2005), pp. 5, 39 et seq.

these weak points was certainly the fact that non-compliance so far has not led to sanctions.

5.6.5 The Public's Participation Possibilities in Planning

Italian practices in the planning of transport and PT service are similar in their formal structure to those of other countries described in this book. For infrastructure projects, there is a multi-stage practice of planning the outline and the details, which provides that the planning documents be made publicly available. As long as funds are needed from the central government – which is almost always the case – the project must also be presented to the Economic Planning Commission.[206] There are also transport development plans on the national and regional levels which, however, do not yet guarantee that the projects they contain will be carried out. Citizen participation in their composition is not provided for.

For LPT, so-called service programmes,[207] produced every three years, are of special importance because they concretely describe the extent and structure of the service and are indeed the foundation for the description of transport services. They are produced by the respective transport authorities (region, province, municipality) which must comply with regional LPT laws.

Participation possibilities and practices come, at present, in various forms. The Lombardian law provides for an advisory work group for transport and infrastructure issues to meet regularly in which providers, political parties, unions and passenger representatives work together. They discuss current issues such as the needs plan or contract specifications. Without any specific legal foundation, there are other commissions and work groups that deal with things like fares, or provide dialogue with various commuter committees (see below). The LPT Law and the transport policy in Lombardy are, compared to the rest of the country, comprehensive and well-developed[208] – the participation possibilities named cannot be expected everywhere. In other regions possibilities are usually limited to discussion rounds without a formal foundation.

It is evident, however, that participation takes place almost exclusively through consumer associations and their representatives. Informal participatory practices which enable laypeople to take part have no tradition in Italy. This applies particularly to the transport sector. Transport is still dominated by technically educated planners who mistrust the 'common sense' of users. The formal participation that occurs only late in the planning process even seems counterproductive in that criticism of experts, at this stage, is often seen as a debasement of their work. Some examples of informal participatory

206 Comitato Interministeriale per la Programmazione Economica (CIPE).

207 Programma Triennale Servizi (PTS).

208 See also the description of transport contracts for the LPRT in Section 3.5.1.4.

practices are, however, to be found in city and district planning and are in part combined with the expectation that such practices can be applied more often in the intermediate term.

There are many accounts of protest actions by dissatisfied customers, which belong to the category of 'civil disobedience'. Examples include the demonstrations and track blockades by commuters affected by service disruptions and delays.[209] Infrastructural arrangements, in addition to the provided participatory practices, could be changes, as the following example from Turin shows:

> It is indeed true that, in reality, participation does already take place, simply in other ways. This is visible in the project of the Turin tramline 4. It runs from north to south through the city and is thus very important. The planners had the idea of upgrading it into a high-speed tram, and thus making it into its own railway trackbed with fences and a relatively large distance between stops. There was meant to be a large P+R lot at the end that would allow one to be in the city in 15 minutes.
>
> Just as soon as these plans were publicized, everyone from the neighbourhood associations, business people, church congregations, etc., set up barricades. One after the other went to the city parliament and to the mayor to advocate for their neighbourhood to have a stop on the line. Then the mayor called the head of the transport operation, who of course got his post on a political ticket too, and ordered him to make a stop in front of the church. The company couldn't give any resistance. So it went on until the line had the station density of a totally normal tramline. The fences along the tracks, of course, came down too because the people wanted to get across the street to the businesses and offices. One doesn't only go shopping on one side of the tracks [this made the fences totally unrealistic]. If you look at it that way, it was citizen participation too! But the project took two years of planning – with a participative approach, one could have had the same result in four months. (report of a Turin city planner)

On the other hand, lack of interest in civic involvement is partly a hindrance to formal as well as informal participation. This can be traced back to a different understanding of politics.[210] In this respect, there are major differences between the North and the South of the country. As a rule of thumb, the citizens' willingness to get involved, as well as the administration's openness, declines as one goes farther south.

209 Legambiente (2006), pp. 15–16.
210 See Gangemi (2004), p. 39.

5.6.6 Interest Representation and Lobby Groups

Passengers' and consumers' associations Associations or groups that represent particularly the interests of PT users are rare in Italy. The only nationally active passengers' association is the Associazione Utenti del Trasporto Pubblico (UTP)[211] founded in 1982 in Milan. The UTP serves first of all to lobby companies and policy-makers, in addition to coordinating regional meetings and occasional conferences. There are also associations of pedestrians and bicyclists, which sometimes take part in PT decision-making and cooperate with other organizations.

General consumers' associations are far more numerous and have far stronger memberships. There are currently 16 such groups that are nationally recognized. With recognition comes the right to limited basic funding and membership in the umbrella group CNCU.[212] There are similar cooperative boards on the regional level. All associations offer their members consultation and support – in disputes with companies, for instance (see Sections 2.5.4 and 5.6.7), and provide user information and political lobbying as well.

The individual associations distinguish themselves through their backgrounds (some are church-based, some affiliated with unions of political parties) and their points of focus. The group with the closest connection to public transport is the Association of Users of Public Services (Assoutenti), with which the UTP closely cooperates. Among other things, the association has conducted several study projects and some surveys of customer satisfaction in LPT, for which partial additional funding was provided by the state. According to Italian law, fines collected by the Antitrust Department must be allocated to consumers, and associations can apply for this funding.

The recognized associations also act as dialogue partners for transport companies and local authorities in formal co-determination practices (see Sections 5.6.4 and 5.6.5). Because their work is generally conducted on a voluntary basis, their wide range of functions is problematic because their activities can lack on-site expertise.

Other organizations Since 1999 the Consumers Forum has provided a mixed consumer-political umbrella group, which is comprised of the larger consumers' associations and some national large-scale enterprises (including Trenitalia) and economic organizations. The association's activities include development of extra-judicial arbitration. Common transport-specific projects have to date only been carried out by individual organizations (see above). Until 2004, Trenitalia together with the associations published the magazine

---

211 http://http://www.geocities.com/utputputp/.

212 Consiglio nazionale consumatori e utenti.

Amicotreno (Friend of the Train), which addressed common passenger issues also from the passengers' point of view.

Corresponding with the gradually developing consciousness of ecological issues,[213] *environmental protection organizations* have played a relatively limited role in transport politics – outside of discussions about pollutant emissions. Since 1998 under the name *treno verde* (the green train), the Legambiente Association together with Trenitalia has conducted measurements of air pollutants and information campaigns in a specially equipped train. The citizens' rights organization Cittadinanzattiva (Active Citizenry) has organized yearly train station parties since 2000 together with train companies.

Commuter committees In contrast to the aforementioned groups, the commuter committees are not fixed entities. They are formless, partially temporary, conglomerations of train customers that gather along one or several lines. Their motivation is usually directly experienced service problems. They work without a charter and with minimal resources (word-of-mouth advertisement, mailing lists, websites), but are at least somewhat integrated into the dialogue between companies, transport authorities and the associations. The exact number of such committees is unknown because to date there has been no national coordination. Eighteen committees are known of in Lombardy, which cover 37 corridors.

5.6.7 Passengers' Rights

Legal foundations According to Article 1218 of the Italian Civic Code, the provider of a service is liable for its potential deficiency, if it cannot prove that it was prevented from providing defect-free service by outside influences. This is true as well for delays in public transport and resulting damages. For processes with low amounts in dispute (up to €1,000), simplified legal action is available with a Justice of the Peace (*giudice di pace*). Because of the fact that the expenses are always high in relation to the possible returns, it appears that processes against transport companies, however, do not take place.

In cases of delay and cancellation, the Trenitalia terms of carriage grant those affected the possibility to either have rides with further trains without additional payment, to stop their trip or to return to their point of departure, in both cases with reimbursement of the fare. There is an alternative option to use other trains and routes than those planned to reach the destination without additional payment, if possible. They thus give reimbursement only for services not delivered, but not as compensation for incurred inconveniences.

Compensation payments on a voluntary basis There are more generous regulations for national and international long-distance trains that reimburse part of the fare in cases of certain delays (Eurostar 25 minutes, intercity 30 minutes, night trains

213 See Potz (1998).

60 minutes), as long as use of the train can be proven with a reservation. With Eurostar there is additional compensation if the climate control fails.

In other forms of LPT, some companies have instituted similar regulations, which are also generally included in their service charters. They range from reimbursement of the fare in cases of transport service disruption over a long period of time to compensation in the form of free tickets in cases of longer delays. Considering the low fares, this only deals with small amounts, which are possibly seen as not being worth the trouble. Compensation for regular customers who have experienced regular delays are just now being offered for LPRT in some regions. (see Section 3.5.1.4).

Possibilities for extra-judicial arbitration of disputes Extra-judicial arbitration of disputes has begun to be offered by various institutions after Law 281/98 significantly expanded possibilities for this practice.[214] The most important authorities in this area are company and consumer associations, chambers of commerce, and municipalities' ombudspersons, as well as companies themselves. This leads to both two- or three-party procedures (with or without the inclusion of a neutral third party).

As a service provided through public commission, PT falls into the realm of responsibility of ombudspersons (*difensore civico*) who help citizens in dealing with the administration. Such positions exist in most regions and, furthermore, in some municipalities and provinces. Their main task is, however, to provide support in problems with the administration in the strictest sense, and as such few complaints about PT are submitted to them.

Consumer associations too offer their members support in submitting complaints. This can come in the form of simple advising or in processing the complaint on behalf of the customer. In some industries, the associations also offer formal arbitration. The foundation for this is a contract between the company and the responsible industry association that regulates the practice and requires both parties to cooperate constructively. In the transport industry, such agreements about arbitration processes can be reached only seldom. Probably the only such service in LPT exists in the Ligurian city La Spezia.

Minimal service in the case of a strike In the case of a strike, there are detailed specifications that can be indirectly counted among passengers' rights. The formal and cultural obstacles to striking are probably lower in Italy than in Central Europe. Striking parties in the transport industry are, however, required to ensure basic service during their strike. This is overseen by a ministerial commission, with whom strikes must be registered beforehand. For PT, this means that this minimal service is not allowed to be denied even on strike days. In local transport, this usually means that trips are made in peak times, in rail travel important regional and select long-distance trains are kept in operation for commuters.

214 Adriano (1998).

5.6.8 Conclusion: A Long Road to Customer Orientation

PT in Italy has obviously been suffering from decades-long neglect, which cannot be quickly overcome. The problem in this is not only providing the industry with the necessary resources. There are 'cultural' factors to be considered as well, which show themselves in transport-political disinterest, in an intermingling with other political goals, in technical- instead of user- or market-oriented planning and in doubts about a change in attitude, even for PT supporters. These factors, or similar ones, are to be found in other countries as well, but seem in Italy to be especially pronounced. The high degree of motorization, too, shows that a change in attitudes may be difficult in this country.

With the state and administrative reforms of the 1990s, major goals were attached to the areas of efficiency, transparency and citizen-orientation. In public transport, too, there was a portion of this: organization and planning procedures began to offer truly comprehensive possibilities for listening to and including passengers.

Because most measures only date back a short time, it is still too early to judge them. In some regions and municipalities, the new possibilities are clearly being used; overall however, it must be attested that many initial approaches have not been carried out, that implementation has been sluggish and that political goals have not yet been fulfilled.

The most visible sign of the effort towards a new orientation to date is the Carta dei Servizi. The widespread scepticism about their actions is all the more astounding in view of the considerable effort that the government has made for the presentation and support of their implementation (see Section 5.6.4). It is obvious that despite their detailed parameters, it has not been possible to communicate the meaning of these measures to the concerned stakeholders on an emotional level.[215] This is, however, decisive for success – for the success of citizen participation in general – and the instruments used to date have only barely begun to achieve it.

5.7 United Kingdom

Anthony D. Smale and Martin Schiefelbusch

5.7.1 Geographical and Political Context

Although the UK is often perceived as one single country from outside, it consists of four main different parts (England, Wales, Scotland and Northern Ireland), each of which has its own history and identity. Furthermore, the spatial structure is very heterogeneous, ranging from the London megalopolis to the very remote and sparsely populated regions in northern Scotland. There

215 See D'Alessandro (2000).

are also still significant differences in terms of economic power and prosperity between London and the southeastern region on the one hand and the old-industrialized cities and rural areas in the north and west of the country.

Despite this, the UK has for a long time been – and in many ways still is – a very centralized country where the lion's share of power is held by the central government in London. Only since 1997 has some delegation (devolution) of powers to regional bodies taken place. Since then, parliamentary bodies have been elected for Scotland, Wales, Greater London and Northern Ireland. These new institutions have also taken over competencies in transport policy and planning. However, the range of powers and institutional setup varies between them.

5.7.2 The Passenger Transport Market

A characteristic feature of British transport policy is that a greater role is given to market forces and private companies. Public transport is considered a commercial activity (with limitations) while most other countries emphasize its role in ensuring accessibility, protecting the environment and social dimension. This difference became most evident in the period of the Thatcher and Major governments (1979–1997), when many of today's structures and planning procedures were set up. However, it can be traced back to older policy traditions – for example, the railways were only nationalized in 1948, later than in most other EU countries. Key policy measures of the 1980s shaping the public transport market are:

- the deregulation and privatization of the express coach market in 1980 and the local bus market in1986;
- the split-up and privatization of the national railway system (completed in early 1997).

As a result, the two main passenger transport markets can be briefly described as follows:

Local and regional public transport Following deregulation of coach services, local bus services in the UK (outside London and Northern Ireland) were freed for competition at the end of 1986. The former National Bus Company was split up and sold, as were many smaller publicly-owned operators. Costs and subsidies were reduced substantially, nevertheless fares rose above inflation in the long run. Patronage sank by about 22 per cent in 10 years, more than was expected from earlier experience.

On the profitable services, on-the-route competition between different operators started with parallel running, schedule-matching, price wars and disruption of networks. In some, especially urban, areas this situation continues until today, although in many cases one company has won the battle.

Furthermore, buying and merging of operators has taken place on a large scale, leading to three major groups each controlling subsidiaries around the country with approximately 60 per cent of the market share.[216] Cases where different operators compete on the same route – as should have been the norm according to deregulation theory – are now limited to a handful of routes.

As not all routes had enough traffic offering to attract private operation or competition, a fraction of services (about 20 per cent) had to be ordered (through tendering) and subsidized by the local authorities. This principle is similar to the controlled competition approach supported by EU transport policy since the 1990s and used in particular in Scandinavia and France. In London the bus network is planned by Transport for London (formerly London Transport) and operations are franchised. In contrast to the deregulated areas, patronage losses did not occur. In Northern Ireland, all public transport is operated by public operators (subsidiaries of the Translink holding) with almost no private sector involvement until today.

In the deregulated areas, the market setup has led to only limited coordination of services. Collaboration across operators was even watched with suspicion by British competition authorities in the early years of deregulation. Furthermore, the system does not foresee any privileges or protection for rail services – both light and heavy – against competition from road-based public transport.

While the New Labour Blair and Brown governments have put new emphasis on an integrated transport system, the basic structure of the industry has remained unchanged. Local authorities have, for example, been given new powers to upgrade bus routes, but this has to take place in partnership with operators and on a case-by-case basis.

Mainline rail The formerly nationalized British Railways was split up and sold between 1993 and 1997. In doing so, the former Conservative government was ahead of its time and went far beyond the separation between infrastructure and operations foreseen by subsequent EU directives. Like in the bus industry, the basic setup of the industry has been maintained by the Blair and Brown governments, although the detailed structure of the railway industry has changed several times, most notably by the bankruptcy of the privatized infrastructure company Railtrack in 2002 and the subsequent return of this part of the industry to public ownership.

At the moment there are 21 passenger train operating companies (TOCs) and several providers of rail freight services operating on Network Rail's infrastructure. In contrast to the bus sector, controlled competition dominates in the passenger rail industry: operators compete for licences to run pre-defined parts of the service (so-called franchises) over a period of seven to 15 years. After some restructuring, the franchises are now awarded by the Department for Transport (for the franchising process, see also Section 4.4.1).

216 White (1997), p. 11.

The government, the Rail Regulator and, last but not least, a statutory passenger watchdog (see below) monitor the performance of the operator against the conditions of the franchise.

The franchising process is very much organized on the national level, but there is a limited involvement of local and national authorities in Scotland, in the big conurbations where suburban services are co-funded locally and in case of Community Railways (see below). Likewise, there is an increasing number of cases of commercial open access operations outside the franchising system.

A far greater degree of integration has thus been maintained for rail services, possibly also as a consequence of the negative experiences in the bus sector. However, TOCs have considerable freedom in their fare and marketing strategies, resulting in an ever greater number of 'special offers' at the expense of transparency.

5.7.3 Organization and Procedures of Public Transport Planning

Following the explanations above, many planning tasks fall entirely into the private operators' realm. Local authorities and politicians have only limited influence, usually through negotiations with the providers or via tendering/ franchising procedures. Passengers, and in particular user organizations, can participate through negotiation and consultations (see below). Looking beyond public transport alone, the main strategy document local authorities use is the Local Transport Plan (LTP). Each authority is obliged by central government to set up a LTP. The same applies to the Passenger Transport Authorities (PTAs), which exist in the seven main conurbations with the main task to coordinate public transport. It is quite common, but not mandatory to consult with citizens when they are drawn up. Usually workshops are held where citizens can voice their concerns and highlight deficits of the transport system. These issues are fed into the following considerations of the authority and the final plan. The evidence submitted with the LTP can support the need for investment measures and the necessary support from central government.[217]

The operators themselves tend to seek information about user needs via market research rather than participation. Following the market economy idea, they are free to decide how they do this, apart from their obligation towards the statutory user organizations. Some bus and rail companies have set up (internal) user advisory boards, but this is not the norm. A major player in the dialogue with bus users is the national organization Bus Users UK who, together with the operators, organizes local hearings (bus user surgeries) for passengers.

---

217 Examples are Wiltshire County Council and Lancashire County Council (for example, http://www.wiltshire.gov.uk/transport-local-plan-consultation-participation.pdf) (accessed January 2007)

5.7.4 The Role of Citizens and Passengers in Public Transport Development

5.7.4.1 Fighting for over a century People tend to believe that rail user groups were unknown before the second half of the twentieth century, but in fact the national campaign group, Railfuture, has discovered a reference to the Southend Season Ticket Holders' Association dated 1894. So it is clear that, over a century ago, there were campaign groups fighting for a better rail service!

Railway history has been punctuated by valiant attempts on the part of central government to devise a master plan which would ensure a consistent level of funding for maintenance of rail infrastructure and service improvement. But the voice of the government Treasury is never far away: 'We simply cannot afford to go on pumping money into the railways.' All too often, the message for passengers is that fares need to rise, poorly-used lines will have to close and the start date for some new piece of railway infrastructure is as far away as ever. Modest improvement schemes which take two to three years to complete in continental Europe need 10 or 20 years just for the planning and consultation stages in the UK.

It is much to their credit that Britain's rail campaigners never give up – the fight continues, year after year, until the argument is won!

5.7.4.2 Local passenger transport groups There must be over 200 rail user groups in Britain. It is difficult to give an accurate figure because new groups are continually being formed while some of the older ones fade away through lack of active members. There are groups that specialize in looking after their local rail stations, making sure the information displays are up to date and the gardens are ablaze with colour. There are others whose main purpose is to campaign for a new station or the reopening of a disused rail line. Common to all groups is an aim to encourage greater use of the railways in Britain.

A new phenomenon is that of the virtual group that springs up on the Internet to protest about an issue affecting large numbers of passengers such as overcrowding. In a very short time, hundreds of people will have added their names to the campaign, particularly if the website is mentioned as part of a newspaper article about the lack of space on trains.

The typical, long-established rail user group will have a strong committee surrounded by a team of perhaps 20 active members drawn from a total membership of a hundred or more. There are user groups in Britain that can boast over one thousand members; the largest is the Friends of the Settle-Carlisle Line with around 3,500 members. The 'armchair' members of a rail user group contribute significantly to the total income so that sufficient sums of money are available to fund newsletters and campaign leaflets, with enough left over for the occasional special event.

The really successful rail user groups will engage with the local community and ensure that their campaigns and activities are relevant to all members of the travelling public. Some examples are:

- on Yorkshire's Penistone Line, passengers enjoy the sound of live music on one of the special 'jazz trains';
- in Cumbria, the Furness Line Action Group has persuaded children from a local school to 'adopt' their station and spend an hour each week on a variety of jobs;
- the Mid Cheshire Rail Users' Association regularly hires a train so that members and guests can enjoy an excursion by rail;
- the Windsor Line Passengers' Association encourages local residents' associations to join, thereby enhancing the relevance of its campaign message;
- the Skipton-East Lancashire Rail Action Partnership organizes public events and conferences to highlight its campaign for line reopening.

Although the number of journeys by bus in Britain outnumbers rail/tram/metro journeys by about three to one, the number of bus user groups remains stubbornly low, at only about 20. There are, however, many groups that represent users of both buses and trains – these are generally called either travellers' associations or passenger transport user groups. More of these composite groups might be seen in the future as articulated buses come to resemble trams and as trams begin to run on railway tracks along with trains.

5.7.4.3 National Representation by Independent Organizations With many local issues to address, rail user groups greatly value their independence. However, many choose to affiliate to a national organization, most notably Railfuture (the Railway Development Society). The RDS styles itself as 'Britain's voice for rail users'. It was formed in 1978 from the merger of two earlier organizations: the Railway Invigoration Society and the Railway Development Association. The name 'Railfuture' was adopted for the twenty-first century, and the group is now constituted as a Community Interest Company.

Railfuture is a founding member of the European Passengers' Federation, and continues to be an active participant of this organization. The EPF lobbies government at the European level and seeks to improve international rail services.

Campaign for Better Transport (formerly Transport 2000) is the name of another well-established, national organization supporting both bus and rail, as well as campaigning for better facilities for cyclists and pedestrians. It has a regional network of campaign groups which bring together people interested in promoting sustainable transport.

Environmentalists are generally supportive of public transport, recognising its significance towards traffic reduction, air quality and efficient use of energy. The principal organization in Britain representing environmentalists is Friends of the Earth (FoE). Compared with other campaign bodies, FoE is truly huge with nearly 100 paid members of staff, around 100,000 supporters and over 200 local groups. FoE UK is part of an international movement with one million supporters across five continents, and transport is always high on their campaign agenda. The support and expertise from FoE is invaluable when mounting a campaign for better public transport; they can advise about everything from the habitat of small mammals to the global effects of climate change.

Since 1986, bus passengers have been represented by the National Federation of Bus Users, now known as Bus Users UK. Over the years, this organization has become adept at holding bus surgeries in towns and cities around Britain. Each surgery is held on a modern, low-floor vehicle parked prominently in the centre of town and passers-by are invited to step on board to meet representatives of the bus company. Volunteers from Bus Users UK are also on the bus to take note of what people say about their bus services, and the same people are invited to a follow-up meeting several weeks later to hear what, if anything, has been done about their complaints. This has proved to be a very effective way to engage with the travelling public.

Table 5.3 Independent national organizations

Was		Now
Railway Development Society	→	Railfuture
National Federation of Bus Users	→	Bus Users' UK
Transport Information and Research Network	→	Association of Community Rail Partnerships
Transport 2000	→	Campaign for Better Transport
Others: Transform Scotland; Friends of the Earth		

Source: author's depiction.

5.7.4.4 Statutory bodies When the railways were nationalized in 1948, the government created a network of Transport Users' Consultative Committees (TUCCs) to look after the interests of passengers. Membership of the committees tended to be drawn from the railway industry, and there was little contact with rail user groups.

Following bus deregulation in 1985 and rail privatization in 1993, the TUCCs evolved into Rail Users' Consultative Committees (RUCCs). People appointed to the RUCCs came from all sections of the community and there was a much improved interaction with the travelling public, including rail user groups.

In 2000, the Central Rail Users' Consultative Committee was reconstituted as the Rail Passengers' Council (RPC) and regional Rail Passenger Committees were set up. These autonomous regional committees took on a variety of campaigns on behalf of rail passengers, much to the annoyance of the Department for Transport!

A few years later, the regional committees were abolished and the RPC was given a new mandate by the government: to concentrate on the gathering of evidence about the experience and concerns of rail passengers across the whole UK network. Thus was born the organization Passenger Focus.

Table 5.4 The evolution of Passenger Focus

Central		Regional
Central Transport Consultative Committee	1948	Transport Users' Consultative Committees
↓	↓	↓
Central Rail Users' Consultative Committee	1994	Rail Users' Consultative Committees
↓	↓	↓
Rail Passengers' Council	2000	Rail Passengers' Committees
↓	↓	↓
Passenger Focus	2006	–

Source: author's depiction.

The two main tasks for Passenger Focus are:

1. to operate the national complaints helpline; and
2. to carry out an annual survey of passenger satisfaction.

Aware of the need to interact with rail passengers outside London, Passenger Focus has appointed 12 Passenger Link Managers, each shadowing one or more of Britain's train operating companies. PLMs ensure that concerns expressed by passengers are communicated to the appropriate train company managers. Rail user groups are encouraged to alert their local PLM to any deterioration in the rail service. In return, PLMs involve the rail user groups in consultations, for example, in advance of a major timetable alteration.

Many rail user groups remain unhappy about the decision to abolish the regional committees, and so new regional partnerships are being set up in their place. The London committee survived the changes and has re-emerged as London TravelWatch, representing both bus and rail users. Transport users' forums outside London are also adopting the TravelWatch title, thus there now exist:

- TravelWatch North West;
- TravelWatch East Midlands, etc.

5.7.4.5 Communities and partnerships One significant battle has now been fought and won. The objective of this battle was to secure the long-term sustainability of rural railway lines, and the group that achieved this was the Transport Research and Information Network (TR&IN). Their idea was to designate rural branch lines as community railways with local management and the support of local rail users.

The UK Department for Transport has now officially endorsed the concept of community rail development. Over 20 railway lines and rail services have been officially designated as community railways, with many more to follow.

On a community railway, the first step is to establish a partnership between the train operating company, Network Rail (the infrastructure company), local authorities and rail user groups along the line. A funding package is then put together to support a paid Community Rail Partnership Officer with a modest budget for rail projects. As the CRP Officer gets to work on schemes to improve the railway environment and encourage people to travel, he or she will rely increasingly on the support of the local rail user groups (to ensure success, several members of Britain's CRPs have travelled through the Channel Tunnel to see how community railways run in Germany). The result is a virtuous circle of improvements and enthusiasm leading to higher passenger numbers and a more cost-effective railway.

TR&IN has evolved to become the Association of Community Rail Partnerships. In May each year, ACoRP hold a Community Rail Day – an excuse for fun and festivities along the community railways of Britain. Another ACoRP event is the Community Rail Awards ceremony each September, when prizes sponsored by the rail industry are presented for achievements such as 'best station garden' and 'best project involving young people'.

It has to be said that some rail user groups in Britain remain sceptical about the value of community railways. They suspect that the aims of Department for Transport are to save money, blame the local community when things go wrong and finally close their branch line altogether. Happily, the experience so far is that community railways are a huge success, with passenger numbers up and active support from all sides of the railway industry. Rail user groups involved in a CRP have a new-found ability to draw on grant funding for minor projects; the partnership enables them to sit around a table with railway managers and talk about service improvements. For the moment, any idea of line closure is unthinkable.

5.7.5 Consumer Protection and Passenger Rights

5.7.5.1 Customer charters and compensation schemes A customer charter (also known as a passenger charter) is a document setting out the level and quality of service a customer can expect, the compensation offered if standards are not maintained and the possibilities to raise comments and complaints. The concept of charters was introduced in the early 1990s by the Conservative Major government with the aim to improve quality and accountability of service providers also in those areas where free market conditions were not (yet) deemed practicable.[218] Therefore the railways and non-deregulated bus operators in London and Northern Ireland had to set up a charter as well.

Today, every TOC must provide a Passengers' Charter as part of the franchise agreement. The TOCs can decide their content and in particular the compensation schemes included, but cannot go below the minimum standards offered by the original British Rail charter. Train operators can of course offer better conditions and have frequently done this because the terms they are prepared to offer are also criterion of the franchising process. Charters have also been introduced by non-franchised companies such as Hull Trains, some light rail systems and by some bus operators who are not obliged to do so.

Usually, the charter refers to clean trains and the maximum time passengers have to wait to buy a ticket or to stand in a train. The following extract is taken from Arriva Train Wales passengers' charter:

> We will clean the outside of our trains every two days and the inside at least once a day. We will not allow a train to enter service without a working toilet. You should usually be able to get a seat. If this is not possible, we will make every effort to ensure that you will not have to stand for more than 20 minutes during your journey.[219]

Regarding compensation for delays and cancellations, Southern, Northern Rail, ScotRail and South Eastern, for example, give vouchers (or cash) equivalent to a minimum value of 50 per cent if the train is delayed by more than one hour (some of them also offer free non-alcoholic drinks). In most cases, customers get the full cost if the train is delayed by more than two hours. Furthermore, season ticket holders can get discounts on renewal if the service has failed to meet punctuality or reliability standards, mostly 5 per cent if one of the standards is not met and 10 per cent if both were not met. Charters used by bus and coach operators are generally less ambitious and do not include compensation for delays and cancellations.

5.7.5.2 Complaints handling Different standards apply also for complaints handling. An elaborate system of standards, monitoring and appeals procedures has been developed for the regulated rail and (partly regulated) bus sectors while formal requirements in the deregulated areas are significantly lower.

For the rail and regulated bus sectors, the main elements are:

---

218 For a discussion of the Charter idea in general see Pinney (1993).

219 http://http://www.arrivatrainswales.co.uk (accessed October 2006).

- *initial complaints*: passengers who wish to complain about a rail service are encouraged in the first instance to take the matter up with the operator providing the service;
- *appeal complaints*: if passengers are not satisfied with the response or do not get an answer within a reasonable time, they can refer their complaint to a watchdog organization[220] which takes the matter up on his/her behalf;
- *TOCs* are obliged to inform their customers about this procedure and to cooperate with the passenger organizations;
- *complaint resolution standards* (mainly response times) are part of the TOCs franchise agreements and are often published in their customer charters.

TOCs are obliged to communicate their complaints statistics to the supervising authority (formerly the Strategic Rail Authority, now the Department for Transport) who publishes them. However, there is no information available on the number of compensation claims because this is considered as commercially sensitive.

With some delay, the deregulated bus industry in England and Wales, has followed with the establishment of the non-statutory Bus Appeals Body (BAB) run jointly by Bus Users UK and the Confederation of Passenger Transport (operator association). In Scotland, the role of BAB has been overtaken by the introduction of the statutory, Bus Users Complaint Tribunal (see also Section 2.5). This leaves only the – relatively limited – light rail systems without an appeals institution.

5.7.6 Conclusions

5.7.6.1 British public transport for the twenty-first century Although there have been some positive outcomes from railway privatization in the 1990s, it is widely acknowledged that there are some fundamental problems with the current organizational structure of Britain's railway. It is a regular complaint that there are now too many parties involved in the decision making process – rail user groups have difficulty finding who is responsible and how things can be changed. Possibly a new structure will emerge but in the meantime, rail user groups are as busy as ever, resolving issues and arguing for more investment in the railways.

With so many user groups representing rail passengers in the UK, it would not be wrong to conclude that the British care passionately about their railways. Across the nation, rail users are working hard to improve their local

---

220 Mainly Passenger Focus and London Travelwatch (see above and Section 4.6.5).

line and ensure that the railways remain an essential element of the transport network for the twenty-first century.

The development of bus and light rail systems is less intensely debated, but it is clear that upgrading and expansion is necessary to make public transport more attractive and able to cope with growing demand. No consensus has yet been reached on whether the current industry structure is able to deliver this. The institutional setup, in particular the division of responsibilities between operators and local authorities, may need to be revised. But the availability of resources appears to be the greater problem. There is no doubt that the 1980s revolutionary changes to the industry have led to significant gains in operator efficiency, but these suffice only to maintain the status quo. At current wage levels of around £7 per hour, the bus industry has for years experienced difficulties in recruiting drivers at least in the prospering parts of the country. The recent expansion of bus services in London which accompanied the introduction of the Congestion Charge has only been possible with additional funding from local government. Even greater difficulties arise when funding of capital investment, for example for light rail systems, is on the agenda. In the absence of regular, transparent public sector support of such measures, a specific solution has to be negotiated for each case.

5.7.6.2 The passenger's position Looking at the different countries presented in this book, British public transport users seem to enjoy a good level of consumer protection and are well integrated in the planning process. Over the last years, quality problems particularly in the railway system, many of which are due to decades of underinvestment, have also highlighted the need for such measures. Clearly, not all problems can be solved in the passengers' interest alone, and the details of charters, compensation schemes and their interpretation are also a matter of dispute between the different parties involved. However, there can be little doubt that these instruments – and the associated monitoring efforts[221] – have delivered tangible benefits in terms of transparency, accountability, empowerment of transport users and operators' management philosophies.[222]

But there are still noticeable differences between the situation in the regulated and deregulated parts of the transport system. These can be explained by historical developments, in particular political decisions, but their adequacy may be disputed. Bus journeys account for the majority of public transport journeys made in the UK, and given the – by and large – failure of on-street competition in this sector, the reasons once put forward for not introducing a system of statutory user representation no longer exist. At last, this seems to have been acknowledged by policy-makers at least as far

221 Monitoring appears at least as important as the schemes themselves – cf. the situation in Italy (Section 5.6.4).

222 Cf. London Regional Passengers Committee (1996).

as user representation is concerned. The devolved governments for Scotland and Wales have decided to set up statutory bodies to represent passengers, and recent proposals for a new Local Transport Bill included the creation of a similar body for English bus users.

5.8 Germany

Martin Schiefelbusch[223]

5.8.1 Political and Geographical Context

With approximately 82 million residents, the Federal Republic of Germany has the largest population of the EU member states. Germany has existed in its current form since the reunification of the German Democratic Republic (GDR) with the Federal Republic on 3 October 1990. Since that time, the coalescence of the two German states with their very different political, economic and social systems has been one of the central societal issues that has shaped all fields of politics. While it has been possible to considerably relieve the deficiencies in infrastructural facilities and maintenance, the economic and social challenges have remained tangible through the years.

Germany is a federal republic with 16 countries, some of which have a longer history to look back on than the country as a whole. The states are solely responsible for many areas of politics and in other areas have a right to participation, while some functions are in the hands of the federal government alone. Local authorities (cities, districts, and municipalities) also enjoy a comparatively large degree of autonomy. Parallel to the division of functional jurisdiction, there are numerous regulations for the financing and sharing of burden between the various political entities.

With its approximately 3.4 million residents, Berlin is the largest city and also forms one of the 16 states. However, on the whole, a polycentric structure with numerous small to mid-size urban centres are typical of the German city system – the Rhine-Ruhr agglomeration, for instance, consists of five cities with more than 500,000 residents alone between Dortmund and Cologne.

During the reconstruction of war damages, there was often special attention paid to providing room for emerging motorization. On the other hand, German cities kept their metropolitan rail transport systems more often than did comparable places in other countries.

223 This section makes use of several sections from the German version of this book which were written by Christian Bodensteiner, Uwe Böhme, Holger Jansen and Mihael Pohar. I should like to thank them for their advice on this section.

5.8.2 The Transport Market and the Organization of PT

5.8.2.1 The formal setting of public transport Public transport services in Germany are, in the far majority of cases, dependent on public subsidies. As opposed to many other countries, however, this has to date hardly been reflected in the legal and organizational framework of PT: the Local Transport Constitution is rather the result of various ideas that, on the one hand, see transport as a normal *commercial* field, and on the other hand, due to its importance as a service of *public interest*,[224] subject it to special regulations that largely exclude it from acting as a free enterprise.

German PT can be structured into three regulatory areas, which make these differing concepts clear:

1. in *local passenger rail transport* (LPRT) – local and regional transport on railways[225] – the classification as a (subsidized) 'service of public interest' is the most clear. With the rail reform of 1996,[226] the organization and financing of this mode of transport was given over to the states. There are no 'commercial services' (which are able to be financed by income from fares alone). Contracts for providing these public services are awarded by public 'transport authorities' in the regions. Underlying the entire service is thus 'controlled competition'; however, despite the common framework, there are still significant differences in its implementation, for instance in the organizational structure of transport authorities, the formulation of their service descriptions, and the periods of transport contracts;
2. *long-distance passenger rail transport*, in contrast, is classified as a *commercial* activity. While in LPRT a range of companies is active, long-distance transport services are provided almost exclusively by Deutsche Bahn AG (German Railways plc). The public has neither the political proxy nor the financial means to subsidize unprofitable transport services. This logic likewise lacks a requirement for a permit

---

224 This translation is used in the following, although the meaning associated with it differs to some extent from the German term *Daseinsvorsorge* (cf. Gegner (2002)).

225 The division between local and long-distance transport is formally determined by distances of more than 50 km and/or one hour of travel time. About 90 per cent of train travel falls into the local transport category. In practice, the division is made by train categories and administrative boundaries, and many 'local' services have route lengths far beyond 50 km.

226 This part of the rail reform first came into effect in 1996 while the transformation of the two national rail companies into DB AG took place already 1994.

or licence for offering particular services, or a demand for coordination between long-distance, local, and other forms of public transport;

3. other *local public transport* (LPT) includes bus, street tram, and underground train transport and is regulated by the Personenbeförderungsgesetz (Passenger Carriage Act (PBefG)). Its basic principles date back to the 1920s; the transport political goal of passing this Act was, firstly, to avoid unregulated competition on the market, and secondly, to protect rail transport – which at the time was lucrative for the state as its owner – from burgeoning competition on the streets. Even though the PBefG does not make any particular statements about the design of rail services, it does provide rail transport companies[227] with a large degree of protection. The law further stipulates that a route licence is required for regular, public transport. In exchange for this exclusive right, there are various requirements stipulated for licence holders, which is essentially meant to guarantee a regulated provision of services according to the parameters of the licence (see below). However, for a long while the PBefG assumed that these services could be provided without subsidies – deficits were balanced by the owners on a voluntary basis. Only in 1996 the possibility to provide non-commercial services was provided for in the Act. Based on the setup for LPRT, these should be awarded by controlled competition as well. To date, however, only a small part of services have been transferred to the public sector because the established companies fear that they will lose their services to competition if bids are invited.[228]

These three fields thus have very different financing and organizational structures. In many areas, the interplay of the various entities in LPT is not clearly regulated and many legal, organizational and financial issues are awaiting clarification.[229] From a practical point of view, this has in part been compensated by the historically close personal and institutional links between local politics and transport companies[230] – if these are willing to cooperate.

5.8.2.2 Public transport providers The service provider's world is dominated by publicly owned companies, even though they are largely organized in a

227 This applies to street tram transport providers as well.

228 Non-commercial provision of services is to be differentiated from commercial provision of transport services, for which direct subsidies are granted by public transport authorities. Most transport services are legally classified as 'commercial' transport services, even though equalization payments are made by the state. For more information on this 'simulated profitability'; see Karl (2002) pp. 18 et seq.

229 See Mietzsch and Oellers (2001), p. 31.

230 See, for example, Projektgruppe Mobilität (2001), p. 22–8, Karl (2002), p. 13, Eichmann (2005).

private legal form. During the course of the rail reform, Deutsche Bundesbahn (German Federal Railways) and Reichsbahn (National Railways in the former GDR) merged to form Deutsche Bahn (DB AG) in 1994 as a public limited company.[231] To date it is fully owned by the federal government, even though its partial privatization has been intensively discussed for some years. Alongside passenger and rail freight transport, DB AG operates numerous bus transport services, particularly in rural areas, and is by far the largest transport company in the country. In recent years, it has also taken a holding in or fully taken over various bus and rail transport providers outside of Germany as well as logistics companies. Inside Germany, DB Netz AG (DB Network plc), a subsidiary of the DB AG holding is responsible for rail infrastructure. This, too, is politically disputed, because it is seen as a hindrance to further competition on the railways.

There are, nevertheless, 300 other rail companies active in Germany, 60[232] of which provide passenger transport. Some of these companies have their own infrastructures, most light railways with a limited coverage. Most of them were founded after the rail reform in order to win contracts through competition, others have existed for a long time. The owners of these companies are local authorities, states, and other transport companies, as well as private interests such as the international concerns Veolia and Arriva.

In the cities and their surrounding areas, local providers generally operate PT. Regional transport in the former Federal Republic is dominated by the previously mentioned federal company DB AG. In the 'new states' (ex GDR) there are predominantly locally owned companies. Most of the c. 6,000 private bus companies in Germany are predominantly active in tourism and charter services; but some operate scheduled routes for their own account or as subcontractors for public companies.

Cooperation between transport providers is a characteristic feature of local and regional public transport in Germany. Starting 1965 in Hamburg, many regions now have a Verkehrsverbund (transport association). These organizations, whose formal background and structure varies across the country, are generally responsible for developing a common fare system for all bus, tram and LPRT services in their area, allocation of revenues, marketing activities and coordination of service planning. In some countries they also have taken over tendering and contract management.[233]

5.8.3 Transport Planning and Participation

According to the federal structure of the Republic, the federal, state, and district/municipal governments are responsible for different functions

231 *Aktiengesellschaft* (AG).

232 Not including heritage, mountain railways, and funiculars.

233 For an older overview on their tasks see Pulcher and Kurth (1996).

in the design and maintenance of the transport network. On the federal level, the Federal Ministry of Transport, Building and Urban Affairs is at the head and has produced, among other things, the Federal Transport Network Plan (Bundesverkehrswegeplan), last passed in 2003, which defines network development for all modes in the federal department's responsibility (motorways, federal highways, federal waterways, and the majority of the rail network). The countries (Bundesländer) are responsible for country roadways and rail-bound local transport (LPRT), among other things; and together with municipalities and companies, they finance a large part of the investment in public transport. Operating subsidies are borne by regional transport authorities for LPRT and local/regional ones for bus and tram services.

Transport services in Germany are designed and defined by various procedures in the three regulatory divisions: in LPRT as well as LPT, public transport authorities produce *local transport plans* as intermediate-term strategy documents, which form the guidelines for the offering and the advised standards. In LPRT and other local public transport where tendering is used, these standards are concretized by tenders, *transport contracts*, and if necessary, renegotiations. Furthermore, *licences* must be applied for or amended for services falling under the Passenger Carriage Act. In long-distance rail transport, there are no licencing requirements of this kind.

5.8.3.1 Participation in the creation of local transport plans With the PBefG and railway reform of 1996, the local transport plan (Nahverkehrsplan, NVP) was instituted as an instrument for the design of LPT services. Since that time, transport authorities in LPT and LPRT have been required to create a local transport plan. The goal is to establish 'the framework for the development of local passenger transport'.[234] A more exact regulation of the design and formation of local transport plans is provided by the LPT laws in each respective state. The content and scope of regulation in former local transport plans show that the discussion about the purpose of this instrument has not yet been concluded. Thus, some local transport plans include specifications about such things as frequency of service and particular timetable data; in others such specifications remain very indeterminate.

The structure of public transport authorities varies from country to country. Usually LPRT authorities have been formed on country or regional level while local authorities are linked to city or district administrations. The transport authority creates the local transport plan, supervises its observance, and has a right to be heard in licencing procedures.

There are legal requirements for the inclusion of third parties in the formulation of a local transport plan, but only a few states' LPT laws make explicit mention of including citizen and/or passenger organizations. The PBefG requires that in the creation of a local transport plan, the transport

234 Section 8 para. 3 PBefG.

authority must cooperate with transport companies (Section 8, para. 3 PBefG) and hear 'representatives of the disabled or disabled persons' advisory councils' (Section 8, para. 3 PBefG). A transport authority can, however, voluntarily include additional groups in the procedure.

5.8.3.2 Invitation of tenders and transport contracts The structuring of contract awards in LPRT is done by the transport authority alone. The procedure is determined by both the official contracting terms for the award of public service contracts and the Public Procurement Code (PPC). There is no provision for public participation in, for example, the framing of the call for tenders. However, in some states the authorities have user advisory boards, which are given the possibility to express their concerns in the preliminary phase. Furthermore, informal dialogue between authorities and user organizations is possible in the work preceding the award of contracts (cf. Section 4.4.1).

Transport contracts that have been made between states and transport companies are generally not available for public viewing. There are thus no exact official reports available about basic financial parameters or agreements about things such as punctuality and according malus payments. It is thus difficult for citizens and user groups to exercise influence. This lack of transparency is also criticized because a significant amount of tax money is used to subsidize public transport.

5.8.3.3 Participation in licencing procedures The second important element in LPT planning is the licencing or authorization procedure. According to the Passenger Carriage Act, transport companies require a licence (*Konzession*) for each scheduled service they operate. The licencing authority is situated on the local or country level and reviews and authorizes licence requests from transport companies. It also authorizes the routing, the number of trips and issues related to fares. In issuing a licence, the licencing authority must comply with the requirements set forward in the local transport plan produced by the transport authority. In the licencing procedure, a series of tests must be performed before the licencing authority grants a transport company the right to provide the requested transport services for a period of up to eight years.[235] The list of bodies to be consulted includes also communities in the catchment area, highway and planning authorities, Chambers of Industry and Commerce, trades unions and industrial unions. But again it is not required to consult individuals or user groups in the licencing procedure.

Changes to existing services, timetables, etc. also have to be approved by the licencing authority. Consultation requirements are the same as for the issue of a new licence, but usually quicker in practice. Again, inclusion of the

235 See Barth et al. (2004), S. 54.

public takes place solely through communication of coming changes in the media – usually shortly before they are carried out.

5.8.3.4 User advisory boards Passenger advisory boards (Fahrgastbeiräte) are the most common form of structured customer participation in German public transport (see Section 4.6.3). Today, about 100 such boards exist in the country, but only a small part of them works on a formal basis such as country transport laws or local statutes. The majority has been set up following decisions of local councils or transport companies' boards. Hence they are usually affiliated to a transport provider or authority, and some do only serve as internal advisors to these bodies while others also work in public.

Regarding their membership, two important models can be distinguished: a 'lay' board consists of passengers who do not belong to any transport political or similar organization. A 'specialist' board, in contrast, is composed exclusively of representatives of such organized groups. There are also mixed forms of both concepts.

5.8.3.5 Informal participation The main activities of passenger groups therefore occur in political lobbying work outside formal consultation procedures (see below and Section 4.6.1). This happens mostly through publicity in appropriate media such as periodicals and trade journals, or newspapers. However, the experiences of train user associations have shown that changes do not necessarily have to occur through public pressure alone. Often, informal cooperation takes place between train user organizations, transport companies, and transport authorities, where ideas and differences of opinion are exchanged. Sometimes it is possible to achieve faster and more effective results in this way than it is through formal practices.

5.8.4 Organizations Representing Passenger Interests

The history of passenger associations in Germany goes back to the early 1980s (cf. Section 4.6.1). The most important organizations representing passenger interests are:

Berliner Fahrgastverband IGEB (Berlin Passengers' Association)[236] The IGEB was founded in 1980 in Berlin-Charlottenburg and can thus be counted as the oldest passenger representation group in Germany. The focus of their work at that time was topics that specifically affected West Berlin, for example, train transit to West Germany, or the situation of the S-Bahn (overground train), which at the time was still operated by Reichsbahn. Today, the IGEB sees

236 Website: http://www.igeb.org (accessed September 2007). The abbreviation is derived from its former name Interessengemeinschaft Eisenbahn Berlin (Railway Interest Syndicate Berlin).

itself as an active representative of public transport passengers' interests in Berlin. The organization is a member of the Deutscher Bahnkundenverband (see below). Their member newspaper is called *Signal*. The IGEB works on a volunteer basis and is independent from transport companies. It currently has c. 170 members.

Fahrgastverband Pro Bahn (Passenger Association 'Pro-Train')[237] Pro Bahn was founded in 1981 in Cologne. The organization was active nationwide in the 1980s, but did not have any country or regional associations of its own. This started to change in 1989. The federal association gradually founded regional associations. In 2006, the association had 4,000 members. The association's magazine is published four times annually and is called 'der Fahrgast'. Pro Bahn sees itself as a passengers' association and representative of interests in relation to transport policy-makers and companies. Pro Bahn is a member of the Bundesverband der Verbraucherzentralen (Federal Association of Consumer Organizations (vzbv)) and in the European Passengers' Federation (EPF). The association works on a volunteer basis.

Deutscher Bahnkundenverband (German Association of Train Customers (DBV))[238] The Deutscher Bahnkundenverband was founded in Berlin in 1990 under the name 'Pro Bahn Hauptverband der DDR' ('Pro-Train Federation of the GDR'). Between 1990 and 1994, the organization cooperated with Pro Bahn in the West German states. Since 1994, however, the association has been working independently under the name Deutscher Bahnkunderverband and adopted a different campaigning strategy. The association has at times taken over rail routes that have been shut down and operated rail transport under their own stewardship with the company Deutsche Regionaleisenbahn (German Regional Railways). The Deutsche Bahnkundenverband has 8,100 members.[239] The association is also a member of the European Passengers' Federation (EPF).

Verkehrsclub Deutschland (Transport Club of Germany (VCD))[240] The VCD was founded in the summer of 1986. Their exemplar was the Verkehrsclub of Switzerland, which was had been founded already 1979. The association's spectrum includes all forms of environmentally friendly transport.

Verbraucherzentrale Bundesverband (Federal Association of Consumer Organizations (vzbv))[241] As a national umbrella association, the vzbv in

237 Website: http://www.pro-bahn.de (accessed August 2007).
238 Website: http://www.bahnkunden.de (accessed September 2007).
239 inquiry with Gerhard J. Curth, President of the Association.
240 Website: http://www.vcd.org (accessed August 2007).
241 Website: http://www.vzbv.de (accessed August 2007).

Berlin combines the work of the country's 16 general consumer organizations and another 25 associations working on consumers' rights. As a nationwide consumers' organization part-funded by the Ministry of Consumer Affairs, the vzbv has 75 full-time employees.

Allianz pro Schiene (Pro-Rail Alliance)[242] The Allianz pro Schiene e.V. emerged as an umbrella organization of 16 associations in the year 2000. The initiative was issued by the chairman of the main railway union Transnet, Norbert Hansen. The Allianz pro Schiene unites associations based on issues of the environment, passengers, consumers, trade unions, and two automobile clubs. The association's main focus is, firstly to do political lobbying, and secondly to offer a discussion platform for their member associations.

There are numerous other groups working predominantly on local or regional level. They are generally committed to one concrete goal, for example to preserve one particular route or make improvements to the timetable, but many of them have contacts with one of the bigger associations as well.

5.8.5 Passengers' Rights

5.8.5.1 Legal foundations for passenger carriage in Germany There is no unified codification of passengers' rights in Germany. The provisions that shape the legal relationship between passenger and transport company have to be extracted from several laws, statutory ordinances, and particularly the terms and conditions of the transport companies (terms of carriage).

German law distinguishes between public rail transport and route-based transport with buses, trolleybuses, and street trams. By way of explanation: these divisions are not intended to differentiate between local and long-distance transport. Local and long-distance trains are treated equally by the law in terms of passengers' rights. Because a more comprehensive discussion of this topic is provided in Section 2.6, only the most important regulations and characteristics are summed up here.

- *Civil Code (Bürgerliches Gesetzbuch, BGB)*: in principle, the provisions in the Civil Code apply to all passenger carriage contracts regardless of the means of transport. In practice, however, the regulations in the Civil Code are rarely referred to. The provisions in the Civil Code are predominantly superseded by the more specific provisions of the statutory ordinances described below.
- *Railway Act (Allgemeines Eisenbahngesetz (AEG))*: the primary legislation for railways is the Allgemeine Eisenbahngesetz or Railway

242 Website: http://www.allianz-pro-schiene.de (accessed August 2007).

Act (AEG)[243] which is the foundation for the Federal Ministry of Transport to pass the Eisenbahnverkehrsordnung, Rail Transport Ordinance (EVO).[244]

- *Rail Transport Ordinance (Eisenbahnverkehrsordnung (EVO)):* the Rail Transport Ordinance applies to carriage with trains that serve the public, as long as no international fares or other regulations determine otherwise. The EVO regulates the legal relationship between passenger and rail company in the style of general company terms and conditions. As a statutory ordinance, however, the EVO is a substantive law. This means particularly that the statutory ordinance's provisions shape the carriage contract between customer and transport as directly as a company's terms and conditions. Because of their legal nature as statutory ordinances they can not, however, be declared ineffective by a court without further proceedings, as would be the case with terms of carriage.

  The EVO was instituted in 1908.[245] Since then there has been a liability exclusion for delay and cancellations. Paragraph 17 EVO of the former version read 'Delay and cancellation do not constitute a claim for compensation'.

  This liability exclusion was first modified on 1 July 2006 when new COTIF regulations about international rail transport were incorporated in German law.[246] Now passengers have had a legal claim to compensation if due to train delay or cancellation, their journey cannot any longer 'be continued on the same day'.[247] The claim comprises 'appropriate costs' that occur for the traveller 'related to overnight accommodation and informing those expecting him' (Section 17, para. 1, p. 2 EVO) and does not apply if the delay or cancellation was due to conditions outside of the rail company's control (Section 17, para. 2 (1), EVO), the fault of the passenger (Section 17, para. 2 (2), EVO), or can be attributed to the behaviour of a third party which the company could not avoid or anticipate despite taking any measures that the situation allowed (Section 17, para. 2 (3), EVO).
- For *other forms of public transport*, there are similar gradations as in the special provisions for train transport in the application of the Passenger

243 Originally from 29 March 1951 (BGBl. I, p. 225), modified 1993 (BGBl. 1993 I p. 2378), last changed with the Act of 27 April 2005 (BGBl. 2005 I 1138).

244 The current EVO was, however, passed in 1938 – before the current AEG was passed.

245 EVO from 23 December 1908, RGBl. 1909, 93.

246 Article 3 (2) of the 'Act regarding the protocol of 3 July 1999 concerning the change to the agreement from 9 May 1980 about international rail transport (COTIF) of 24 August 2002 (BGBl. II p. 2140).

247 For the issue of whether the norm should be interpreted such that there are no further limitations of liability, see Bitter (2005) and Pohar (2006), p. 342.

Carriage Act (PBefG),[248] the Ordinance for General Terms of Carriage for Street Tram and Omnibus Transport, and Route-Transport with Automobiles (VO AllgBefBed),[249] and finally the special terms of carriage (Besondere Beförderungsbedingungen, BBB) of individual transport companies. For claims in cases of delay or cancellation of a mode of transport, Section 16 VO AllgBefBed stipulates that there is no claim for compensation in cases of deviations from timetables due to obstructions of transportation, service disturbances or discontinuance, as well as shortage of space and thus no guarantees are given for meeting connections.

- *Conditions of Carriage*: regardless of the mode of transport, the primary source for customers' rights is the general terms and conditions of the respective company. These terms and conditions are effective principally by virtue of contractual agreements. The passenger thus has an enforceable contractual right to receive the compensatory services promised in the terms and conditions, as long as the necessary preconditions are fulfilled. Terms of carriage are created by the transport company and thus give a one-sided determination of the relationship to the passenger. In other words, it is in the hands of the transport company alone to allow or refuse rights to the customer. Customers are not, however, totally helpless in the limitations of the terms of carriage. If they infringe upon the compulsory norms of the Civil Code, they are ineffective and cannot be applied.
- DB AG's terms of carriage offer compensation in cases of delay in the following way. If the destination is reached by trains of the product class ICE or IC/EC more than 60 minutes late, travellers receive compensation in the amount of 20 per cent of the value of the ticket, and for round-trip journeys 20 per cent of half the ticket value. This also applies if the delay can be attributed to cancellation of a train within a chain of travel connections. The amount of compensation must be at least €5 and is rounded to an amount divisible by ten cents. The passenger does not, however, receive cash, but a voucher that is valid for one year.

5.8.5.2 Customer charters and service guarantees DB AG's abovementioned compensation scheme is also part of the company's 'Customer Charter' (Kundencharta), officially launched on 1 October 2004. Many local transport operators in Germany have introduced similar service guarantees since the late

248 Of 8 August 1990 (BGBl. I, p. 1680), last changed with the act of 19 July 2002 (BGBl. I, p. 2691).

249 Of 30 June 1989 (BGBl. I, p. 1273), last changed with the act of 12 December 2001 (BGBl. I, p. 3765).

1990s.[250] The transport association of Rhine-Ruhr (VRR), for example, offers a Mobility Guarantee. According to this guarantee, taxi costs are reimbursed if a means of transport departs with a delay of more than 20 minutes and if there is no alternate connection. Similarly, the transport association of Rhine-Sieg (VRS) offers, in the context of a mobility guarantee, to reimburse taxi or even ICE costs in cases of departure delays of more than 20 minutes. In cases of more than 20-minute delays, the BVG in Berlin offers passengers a free ticket during the day, and during the night reimburses taxi costs of up to €25.

However, *ex gratia* payment regulations and voluntary declarations of commitment from transport companies have to be distinguished from legally binding terms of carriage and the other regulations described in the previous section. As long as they are not identical with the terms of carriage, charters and guarantees are voluntary declarations of commitment and fundamentally not legally binding.

5.8.5.3 Settlement of disputes The legal framework described above largely prevents passengers from seeking legal action in case of disputes about service quality and cancellations. Dissatisfied users can complain to local councillors, parliamentary petition boards or seek to attract media attention and have occasionally been successful in the past in getting compensation or a better response to their queries. User associations and advisory boards can take up complaints as well, but again without formal arrangements or backing to do so. They usually also do not have the resources to deal with complaints systematically.

The first arbitration body in German public transport was set up in 2001. Since then, several others have been established (cf. Section 2.5.4), but until late 2007 only three of the 16 states had set up alternative dispute resolution offices for local public transport.

5.8.5.4 Discussions about reform in Germany Because liability for train delay and cancellation was completely excluded in Germany according to Section 17 EVO up until 30 June 2006, passenger associations and consumers' rights organizations have long been calling for passengers' rights in national passenger rail transport to be strengthened legally, or to dismantle the out-of-date liability privileging of rail companies. The rise of train delays and cancellations due to economizing measures by the DB management in recent years have intensified the discussion.[251] Since at least the beginning of 2002,

250 For details see Sec. 2.4 and EUSG-Consortium (2006).

251 For more extensive information see Bodack (2004), pp. 7 et seq. and the *Süddeutsche Zeitung* from 29 September 2003, 'Bahn selbst Schuld an Verspätungen – Interner Prüfungsbericht verweist auf viele Mängel' (Train Company Itself Accountable for Delays – Internal Audit Shows Many Deficiencies'). Also, derFahrgast 1/2004, 21 et seq.

passengers' rights have been on the agenda in federal and country politics. Several proposals have been made on how to improve consumer protection, but no agreement had been reached until spring 2008. Following the adoption of EU Regulation 1371/2007 on passenger rights and obligations in the rail sector, Germany will have to make changes to its legal framework, but it remains to be seen whether this will go beyond what is required by the EU.

5.8.6 Conclusion: Does Voluntary Engagement Suffice?

In the context of Europe at large, Germany is in the mid-range in terms of the possibilities for passenger associations' work. Countries such as the UK and The Netherlands offer better possibilities, in particular regarding formal, guaranteed consultation and participation possibilities and support for user organizations' work. In Germany, there is no explicitly regulated and formalized participation in the local transport plan development or licencing procedures. If citizen participation is compared to other fields of planning, for example building and development legislation, the institutionalized possibilities for cooperation are strikingly small.

Compared to this, it is certainly positive that Germany has a number of associations campaigning for passenger interests who, despite of limited resources, have been active for nearly three decades. They were often able to develop a good informal dialogue with the providers and influence transport planning and policy decisions on several occasions.

Consumer protection in Germany is also characterized by quite a number of positive informal initiatives – in particular service guarantees – but relatively poor legally binding standards. The question arises if this situation is beneficial for transport users – and indeed the transport industry as a whole – in the long term. The fact that passenger satisfaction is consistently low in Germany both compared to other sectors and to other countries[252] indicates that there is still some work to be done.

5.9 Outlook: Different Solutions for Similar Problems

Holger Jansen

How do transport service providers involve their customers, the passengers, in issues of future development? And how do passengers' rights work if things do not go as the company promised? These were the central questions of this chapter. It is no wonder that the situation in seven European countries should be different. Certain structures are similar in many countries. But – perhaps

252 See, for example, Blümel (2004), p. 7, European Commission (2004a) and vorm Walde (1997).

because national backgrounds and cultural differences play a role – there are significant differences.

In most countries, political and corporate responsibilities have become separate. Only in the Czech Republic has this course not yet been completed. Instead, what happens is that policy defines the framework, and the companies provide the services. Differences exist in terms of competition. In the UK and Sweden, portions of the LPT market have been tendered for decades and thus put into a competitive sphere. In the deregulated areas of British transport, several companies can be competing for customers on the same route. Austria and the Czech Republic are at the very beginning of this development.

All of the countries have a similar background in terms of politics. The Ministry of Transport oversees transport on the national level, and thus address particularly long-distance transport. Local and regional transport, in contrast, are organized by the regions, departments, or provinces. This level in turn entrusts transport authorities or other organizations with operative implementation.

On the regional level, passengers too have the chance to be intensively involved in discussion processes through advisory boards, discussion platforms or other means. On this level, legal claim to participation often turns out to be quite limited. Legal claims to participation exist only in protracted or extensive planning for major projects. In everyday business, such as changes to timetables, the passengers or the associations representing customer interests are dependent on the good will of the company or the organization responsible.

In all of the countries, there are special organizations that address the rights of bus and train customers. These organizations' national situations differ in terms of the size of their membership, their employment opportunities, their structure and their financial means. In the United Kingdom, national organizations receive public funding to fulfil these tasks.

In other countries, participants involved on an voluntary basis or a very small staff of full-time employees head the associations. The countries' general consumer associations are also involved in public transport. They have the advantage that they are for the most part state-funded and can thus do long-term work. Public transport is, however, only a small part of their work. Trade associations for the representation of passenger interests often have a better understanding of the industry than general consumer associations.

Almost all countries have instituted an arbitration board or something similar for alternative resolution of disputes in public transport (see above, Section 2.5). In 2007, a new arbitration board was introduced in Austria for rail transport. In the Netherlands, there has been a similar organization since the 1990s. In France a *médiateur* can be engaged in some areas in case a passenger is dissatisfied with the transport company's response. These arbitration boards should be lead by an advisory board on which service providers, customer associations and other stakeholders are represented.

This presents the important opportunity to recognize and intercept structural deficiencies through an examination of individual grievances.

Passengers' rights in each country are closely connected to the settlement of disputes. Claims provided by laws, ordinances or companies' general terms of carriage are limited in most countries, most provide only further transport with the least possible delay. There have, however, in many countries been positive developments for the customers. In Britain, for example, there are minimum standards for national terms of carriage. Many companies in the United Kingdom have voluntarily exceeded these. In Sweden, 95 per cent of all trips are covered by a service guarantee. Under certain conditions, if there is a delay of 20 minutes or more, the customer can use a taxi.

Many transport companies or transport authorities voluntarily offer these requirements in the form of a charter or service guarantee. Only in the Czech Republic and Austria does this form of quality guarantee play almost no role.

There are significant differences in the countries concerning the various dialogue and participation possibilities. In The Netherlands, there are publicly supported discussion forums in which associations, companies and political decision-makers discuss public transport. In the other countries, many associations are dependent on the good will of their respective providers. In the Czech Republic, there is currently no institutionalized participation. In the foreground here are improved quality in rolling stock and a resulting stabilization of the market share in public transport.

What are the particular strengths of the individual countries? In the United Kingdom this is certainly the public financing of national passengers' 'watchdog' organizations. In Sweden, consumer protection has a long tradition, beginning in 1809. In The Netherlands, the legal framework for citizen participation was set in 2000. In France, many companies institute certification processes to guarantee quality.

One thing, however, is the same in all countries. Public transport is always dependant on public co-financing. The level of cost-coverage differs significantly from country to country. Public funding is, however, indispensable for functioning LPT. Finally, it must be concluded that in many countries, due to external influences, the transport market still has considerable change processes to undergo. All stakeholders should make an effort to maintain the right tension between economic necessities on the one hand and active customer orientation on the other. Citizens and the associations that represent them should therefore be involved in these processes.

Chapter 6
The Future of Passengers' Rights and Passenger Participation

Hans-Liudger Dienel[1]

This book has presented a wide array of various techniques and concepts, ranging from passenger participation in the development and design of transport services to enforceable passengers' rights. Both areas overlap and share a common overall objective: to strengthen public transport.

Why have public transport experts not discussed these issues for such a long time and why are they now beginning to become more and more prominent? There are evident historical reasons for this. Firstly, the legal framework for railways and its specific form of customer-orientation developed in times when it actually had a monopoly as a transport system in many areas. As the rise of motorized individual transport put pressure on PT, it survived in Europe thanks to public support and governmental protection. The safeguarding of this support and of legal sheltering had a central meaning for public transport services in the post-war decades. They were also able to advance by leaps and bounds through the decades-long fixing of fare rates and franchising in goods transport, which held road-based competition at bay. The limitation of enforceable passengers' rights was also among these factors. This was certainly a justifiable effort to protect an ailing public transport industry against customers' demands in order to ensure that it would survive at all.[2]

Indeed, in the long term this was politically disastrous for the transport companies, which slowly but permanently came to be seen as shrinking or in retreat. Its employees, in contrast to earlier times, had a crestfallen self-awareness, and were additionally demoralized by increasingly obsolescent technology. These were times when some employees of local public transport systems did no longer used the services themselves and even viewed their customers with a certain disdain. A growing proportion of the companies' revenues were no longer earned from passengers, but in the form of subsidies from public funds.

Indeed, a widespread change has set in since the 1980s. Strengthened by new, ecological arguments for the construction of public transport systems,

1 The author would like to thank Martin Schiefelbusch for his helpful suggestions.

2 Dienel (2004, 2007).

this change took the shape of investment in new technology, growth and offensive customer-orientation. It began in long-distance transport with the targeted construction of inter-city transport and, in the 1980s, a new marketing approach – Germans remember Hemjö Klein's 'pink elephants' (*Rosarote Elefanten*)[3] and the BahnCard[4] – and later the new high-speed routes.[5] Large cities invested in new metropolitan transport systems. Until the end of the century, new trains and offerings in regional transport began to emerge. Transport companies and their employees today are not only significantly more customer-oriented than 20 years ago, they are also once more confident about the future and the growth prospects of their service, and therefore are more open to passengers' rights and passenger participation.

In what direction will and should the development of passengers' rights and passenger participation in Germany and Europe continue? In short: passengers' rights and passenger participation will continue to increase in importance. In conclusion, this chapter suggests three main reasons for this.

The European Commission as a Motor for Passengers' Rights

Recent studies document how the European Commission has, for the last few years, been systematically and consistently advancing the development of passengers' rights.[6] Of particular mention here are the Directorate-General Consumer Protection and Peter Faross' department in the Directorate-General Energy and Transport. Around the turn of the millennium, the Commission began to strengthen its work in the area of passengers' rights starting with a concentration on air transport. In the meantime it has also taken on train, bus and ferry transport and, moreover, has supported cooperation with passenger and consumer organizations on the European level.

The improvement of passengers' rights has been a tough struggle over details. The committed efforts on the working level clearly have the Commission's backing, and these efforts fit almost ideally into the overall concept of strengthening citizens' rights, with which the European Commission has been trying to make their citizen-oriented approach visible and tangible in recent years. Particularly since the European Constitution was defeated in the vote in 2005 in France and The Netherlands, two core EU

3 The pink elephants were used in the railway's publicity material to advertise new bargain fare offers.

4 National rail discount card for which user organizations like VCD campaigned also for a long time. It was finally introduced as an offer for the general public in 1992 (group-specific discount passes had already existed for some time).

5 Hemjö Klein was responsible for marketing on the executive board of Deutsche Bahn starting in 1982. See Albus and Heine (2002).

6 See http://ec.europa.eu/consumers/topics/facts_en.htm (accessed March 2008).

founding countries, the topics of citizen-orientation and citizens' rights have been significantly revalued in the Commission. The programme 'Plan D' (Plan Democracy), launched by Commission Vice President Margot Wallström, is meant to drastically strengthen European citizens and thus lead to rights for cooperation in the formation of Europe. Passengers' rights profit from this concept as well.

The next step is legal safeguarding of passenger *participation* in public transport as well. These efforts are still just beginning. On the establishment level, to date there are still mainly only voluntary commitments by transport companies, who see an improved image, new solutions, and in particular more passengers. Indeed there is a recognizable trend: the movement is in the direction of strengthened cooperation with customers in the services of public transport.

Governance of Infrastructures as a Motor for Passenger Participation

Passengers' rights and passenger participation exist in the context of a fundamental reconstruction of the infrastructures and service utilities available to the public. This reconstruction is moving away from services for which public governance alone has responsibility and control, and towards services that are designed and maintained together by the state, economic partners, and citizen engagement. In this context, the importance of passenger participation will continue to increase because people who take part want to be heard and to take part in decision-making as well.

LPT is affirmed as an important public asset, and most people fundamentally welcome its public co-financing. Other public assets, such as those in the area of culture, are under more pressure. Nevertheless, the companies have been unsuccessful in gaining market shares on a broader front. In contrast, especially in rural regions, demand has been decreasing for years, and it seems increasingly unlikely that it will continue to be possible to finance existing services. This is one of many examples of the fact that the state has seemingly overextended itself in infrastructure. In this situation, which is particularly dramatic in remote areas with shrinking populations, there are basically two alternative strategies: *deconstruction* of infrastructures, such as shutting down and thinning out transport routes, closing swimming pools, theatres, and schools – or *reallocation* of the burden through the integration of partners in industry and civil society. This has been the case for many regional infrastructures in rural areas. This principle of *civil-societal infrastructure* can be defined in the following manner: things that the state has previously done alone start to be done together by government, industry, and the third sector in trisectoral cooperation. In increasingly more areas, citizen engagement in association with the engagement of local companies is securing and supplementing the municipal, local, or regional infrastructures as a public asset. In this way, the quality of infrastructural services improves at the same time.

An example of this kind of development is the rise of citizens' buses. Citizens' buses are a civic effort to provide mobility for people where classic public transport is not (any longer) available. The first citizens' bus in Germany was created in North Rhine-Westphalia as a supplement to eroding PT. Since then, citizens' bus organizations have started all over Germany and service over 50,000 passengers per year.[7]

Participation for a Better Public Transport

Competition is generally seen as an impetus for customer-orientation on the part of service providers; but this rarely occurs in PT, as described in Chapter 1. The reforms of recent years have, however, sought to structure transport with an orientation toward competition. Though they only indirectly make competition for the 'passenger as end customer' their goal, developments to date have shown positive effects on customer-orientation. Whether this is a lasting development or purely carried out as a cost-minimizing strategy remains to be seen.

One precondition for customer-oriented service is that companies are aware of customers' expectations, and that they are given sufficient attention in the planning process. Today, however, as the 'purchaser' of public transport systems, transport authorities are the most important 'customers' for many transport companies. They are carefully monitored, cared for, and served. This is good in its own way. Indeed if a transport company does not want to live exclusively on a diminishing number of 'captives', that is, users who have no alternative to public transport systems, but wants instead to draw more voluntary customers, it should strengthen passengers' rights and use passengers purposefully to improve their service. Thus, customer-orientation remains an important area for future development in local transport.

In a wonderful lecture at the conference 'More Customers through More Participation?', Dr Otmar Lell of the German Federal Consumer Advice Centre compared the intensity that European automobile builders applied to research for the optimization of the 'SLAM' of car doors with the lacking empathy for train passengers.[8] In public transport, it is particularly apparent how severely transport science's tradition of a technological-operational-economic approach can shape the results of planning processes.

There is a long way to go in this area; it is, however, a worthwhile process. Not only the automobile industry, but air transport as well has been forced

7 See www.pro-buergerbus-nrw.de, www.buergerbus-gransee.de, www.mobilkult.de (accessed March 2008).

8 Lell (2005).

into customer-orientation from the start. For this reason, the structure of aircraft cabins is in many respects more customer-friendly and better suited to customer needs, even though the spatial conditions in trains are actually more convenient. This includes the leathery, high seats, which afford more privacy and comfort than many seats on the train. Airline companies also have more experience with the growing demand for 'delaytainment', that is, creative, customer-friendly structuring of waiting time as an event.[9] This book ends with this thought: only when public transport systems can compete on the empathetic level with individual transport systems will they have the possibility of developing the market of voluntary passengers. Strengthening passengers' rights and passenger participation are two ideal approaches to achieving this goal.

9 Sauter-Servaes and Rammler (2002).

Bibliography

Academy for Technological Impact Assessment in Baden-Württemberg (TA-Akademie) (ed.) (2002), *Empfehlungen der Bürgerforen im Rahmen des Projekts Mobilist. Bürgergutachten. Arbeitsbericht Januar 2002* (Stuttgart: Author's Edition).

Adriano, F. (1998), 'Se, un giorno, gli utenti potranno difendersi da soli', *Trasporti pubblici* 11, pp. 12–14.

Anon. (1997), 'Confident Souter Gives Upbeat Message of Prospects for Operators', *Transit* 13:11, pp. 8–9.

Anon. (2004), 'Schweizer Verhältnisse in Hamm', *Ruhrschiene* 3, pp. 1, 3.

Anon. (2006), 'Treni sporchi e affollati', Altroconsumo January, pp. 21–3.

Anon. (2007), 'Wissen, mit wem man reden muss', Dr Otto Wiesheu, Vorstand bei der Deutschen Bahn und politikerprobt, vertritt die Interessen des Konzerns in der Welt und sucht nach Verbündeten, *DB Welt* 3, p. 3.

Apel, D. (1992), *Verkehrskonzepte in europäischen Städten* (Berlin: difu).

Apel, H. (ed.) (1998), *Wege zur Zukunftsfähigkeit – ein Methodenhandbuch, Arbeitshilfen für Selbsthilfe- und Bürgerinititativen Nr. 19* (Bonn: Stiftung Mitarbeit).

Arbeitskreis 'Beteiligungsverfahren' der Fachgruppe, Forum Mensch und Verkehr' in der Vereinigung der Stadt-, Regional- und Landesplanung SRL e.V. (ed.) (1998), *Kooperative Beteiligungsverfahren in der kommunalen Verkehrsplanung, Anregungen zum Tätigwerden* (SRL Schriftenreihe No. 44, Berlin: SRL e.V.).

Assoutenti Lombardia u.a. (2005), *Tra qualità percepita e qualità offerta – Proposte per servizi a livello delle migliori esperienze Europee* (Milan: Author's Edition).

Authority of Health and the Environment, Hamburg (ed.) (2001), *Kursbuch Umwelt*, http://www. hamburg.de/Behoerden/Umweltbehoerde/kursbuch/kursbuch.htm.

Automobile Club Italia (ACI) and Istituto Eurispes (2006), *Rapporto sulla qualità della mobilità nelle province italiane. Sintesi per la stampa* (Rome: Author's Edition).

Baer, R. and Hill, D.J. (1994), 'Excuse Making: A Prevalent Company Response to Complaints?', *Journal of Consumer Satisfaction, Dissatisfaction and Complaining Behavior*.

BAG-SPNV (2006), BAG-SPNV, *Fahrgastrechte – gut, dass der Zug ins Rollen kommt!*, http://www.bag-spnv.de/index_.htm, accessed June 2007.

Balducci, A. and Calvaresi, C. (2000), 'A Case Study: The Milan City Region', in *Fachgebiet Europäische Raumplaung Universität Dortmund: Italien – Raumplanung und Stadtpolitik* (Dortmund: Author's Edition), pp. 71–9.

Barlow, J. and Møller, C. (1996), *Eine Beschwerde ist ein Geschenk (Der Kunde als Consultant)* (Vienna: Ueberreuter Verlag.).

Barnes, M. (1999), 'Researching Public Participation', *Local Government Studies* 25:4 (Winter), pp. 60–75.

Barth, S., Baumeister, H. and Berschin, F. (eds) (2004), *Recht des öffentlichen Nahverkehrs – Praxis Handbuch* (RD. 54, Munich, loose-leaf-collection).

BASIC-Konsortium (2005), *The Provision of Basic Services in Liberalized Markets (BASIC)* (Deliverable 8 – Final Scientific Report. Vienna: ICCR).

Baumeister, H. et al. (2005), 'Kundengarantien und Fahrgastrechte in Verkehrsträgen des ÖPNV. Eine neue Qualitätssteigerung durch Servicegarantien und Kulanzleistungen', *Verkehr und Technik* 12, pp. 485–9.

Becker, J., Behrens, H. and Hollborn, S. (2003), 'Qualität von Nahverkehrsleistungen. Die Bedeutung der neuen DIN EN 13816,' *Internationales Verkehrswesen* 1+2, pp. 30–34.

Beckmann, J. and Keck, G. (1999), *Beteiligungsverfahren in Theorie und Anwendung* (Stuttgart: Author's Edition).

Beckmann, K.J. (2000), 'Bewertungsverfahren weiter auf dem bisherigen Weg? – Erfordernisse eines Methoden Mix', in Martens, S. and Brenner,J. (eds), *Akademie für Technikfolgenabschätzung in Baden-Württemberg, Arbeitsbericht 182 'Bewertungsverfahren im Verkehrswesen: Rechenstift gegen Argumente?'* (Stuttgart), pp. 37–73.

Behörde für Umwelt und Gesundheit Hamburg (ed.) (2001), 'Kursbuch Umwelt' (inhttp://www.hamburg.de/Behoerden/Umweltbehoerde/kursbuch/kursbuch.htm).

Berschin, F. (2000), *Daseinsvorsorge durch Wettbewerb (der öffentliche Verkehr zu Lande im Markt)* (Heidelberg: Inauguraldissertation zur Erlangung der Doktorwürde der Juristischen Fakultät der Ruprecht-Karls Universität [unpublished]).

Bischoff, A., Selle, K. and Sinning, H. (1995), 'Informieren, Beteiligen, Kooperieren – Kommunikation in Planungsprozessen', *Schriftenreihe Kommunikation im Planungsprozess*, Vol. 1 (Dortmund: Dortmunder Vertrieg für Bau- und Planungsliteratur).

Bitter, G. (n.d.), *Haftung der Deutschen Bahn für Zugverspätungen und fehlerhafte Fahrgastinformationen* (in NZV 2005, 553).

Björlin-Lidén, S. (2004), *The Role of Service Guarantees in Managing Services* (Karlstad: Universitetstryckeriet Karlstad).

Björlin-Lidén, S. (2005), 'Service Guarantees – A Comparison across Sectors', presentation at the conference *More Customers through More Participation?*, 29–30 September, Berlin.

Blanckenburg, C. von, Böhm, B., Dienel, H.-L. and Legewie, H. (2005), *Leitfaden für interdisziplinäre Forschergruppen: Projekte initiieren – Zusammenarbeit gestalten* (Munich: Franz Steiner Verlag).

Blümel, H. (2004), *Mobilitätsdienstleister ohne Kunden – Kundenorientierung im Öffentlichen Nahverkehr* (Berlin: WZB discussion papers SP III 2004–109).

Bockholt, T. and Probst, G. (2003), 'Kundengarantien im ÖPNV', *Der Nahverkehr* 3, pp. 23–30.

Bockholt, T. and Jugelt, R. (2005), *Fahrgastorientierung durch Kundengarantien. Chance für eine neue Qualitätssteuerung*: http://www.newstix.de/downloads/fachbericht1623.pdf (accessed January 2006).

Bodack, K.-D. (2004), 'Im Schleudersitz als City-Hopper – verlieren wir die Kultur des Bahnreisens?', *der Fahrgast* 1, pp. 7–14.

Bodensteiner, C. (2005), *Bürgerbeteiligung in der Verkehrsplanung. Untersuchung der Bürgerbeteiligung bei der Einführung des Metrokonzeptes in Berlin* (Bodensteiner).

Bodensteiner, C. (2006), 'Die Bürgerbeteiligung bei der Einführung des Metrokonzeptes in Berlin', Becker, H.-J. and Kracker, E. (eds), *Beiträge aus der Verkehrsforschung* (Berlin: Universitätsverlag der TU Berlin), pp. 35–52.

Böhm, B. (2006), *Vertrauensvolle Verständigung – Basis interdisziplinärer Projektarbeit* (Munich: Franz Steiner Verlag).

Böhme, I., Krause, J. and Volpert, M. (1998), *Vereinigung für Stadt-, Regional- und Landesplanung SRL e.V., Kooperative Beteiligungsverfahren in der kommunalen Verkehrsplanung – Anregungen zum Tätigwerden* (Berlin: SRL-Schriftenreihe).

Bogumil, J., Holtkamp, L. et al. (2001), *Verwaltung auf Augenhöhe. Strategie und Praxis kundenorientierter Dienstleistungspolitik* (Berlin: Edition Sigma).

Borth, B.-O. et al. (2004), *Beschwerdezufriedenheit und Kundenloyalität im Dienstleistungsbereich (Kausalanalysen unter Berücksichtigung moderierender Effekte)* (Wiesbaden: Deutscher Universitätsverlag).

Bracher, T., Eichmann, V. and Kühn, G. (2004), *ÖPNV im Wettbewerb, Management-Planspiel in der Region Berlin* (Berlin: difu).

Bramezza, I. (1998), 'Italy', van den Berg, L. et al. (eds), *National Urban Policies in the European Union* (Aldershot: Ashgate), pp. 224–45.

Brinkerhoff, D. and Goldsmith, A. (2000), *Macroeconomic Policy, Poverty Reduction Strategy and Participation* (New York); http://www.worldbank.org/participation/web/webfiles/macrosynthesis.htm (accessed March 2005).

Bröer, G. (1994), 'Volkes Stimme gegen Betriebsblindheit', *fairkehr* 6, pp. 44–6.

Brög, W. (2000), 'Switching to Public Transport', presentation at UITP Asia-Pacific Conference, 10–13 October, Melbourne.

Brög, W. and Erl, E. (2004), 'Just Do It! Wegweiser für Verhaltensänderungen', presentation at Fachtagung anlässlich des 60. Geburtstages von o. Univ. Prof. Gerd Sammer, Universität für Bodenkultur Wien, 26 March).

Bruhn, M. (1996), *Qualitätsmanagement für Dienstleistungen. Grundlagen, Konzepte, Methoden* (Heidelberg: Springer).

Bundesamt für Verkehr (BAV) (2004), Pressemitteilungen vom 7., 16., 19., 20., April, 11. May 2004 and supplementary oral information.

Bundesministerium für Verkehr, Bau- und Wohnungswesen (BMVBW) (ed.) (1999), *Planungshandbuch für den öffentlichen Personennahverkehr in der Fläche* (Bonn: Fach-Media-Service-Verlag).

Bundesministerium für Verkehr, Bau- und Wohnungswesen (BMVBW) (2004), *Mobilität in Deutschland 2002 – Ergebnisbericht* (Berlin: unpublished).

Bundesverband Bürgerinitiativen Umweltschutz (BBU) (1996), *Bürgerinnen und Bürger initiieren Zukunft, Handbuch für Bürgerinitiativen* (BBU-Argumente II/96, Bonn: Author's Edition).

Bundesverband Bürgerschaftliches Engagement (bbe), (n.d.), *Glossar europäischer Netzwerke*: http://www/b-b-e.de, accessed March 2005.

Burger, K. (2006), 'Ein Mann gegen 15 000', *Die Zeit*, 17 August.

Burow, O.A., Marschall, S., Schulte, D. and Taubken, N. (2002), *Aktuelle und bewährte Konzepte zur Gestaltung und Moderation von Großgruppenveranstaltungen – Ein Wegweiser* (unpublished).

Buschmann, V. and Grunwald, M. (1998), 'Fahrgäste als Qualitätstester', *Der Nahverkeh*r 11, pp. 26–30.

Callies, H. (1959/1991), 'Praktische Psychologie im modernen Großstadt-Massenverkehr', presentation at DVWG Rhein-Ruhr 1959, published 1991 in *Verkehrszeichen* 2, pp. 18–23.

Carminucci, C. et al. (2005), *'Avanti c'è posto?' – Report annuale ASSTRA – ISFORT sulla mobilità urbana: I bisogni dei cittadini, le risposte della mobilità* (Rome: ISFORT).

Cartledge, J. (2005), 'Watchdogs. The Example of LTUC and the British System of Institutionalized Representation', presentation at the conference *More Customers through More Participation?*, 29–30 September, Berlin.

Central Rail Users Consultative Committee (CRUCC), London Regional Passengers' Committee, Office of the Rail Regulator and Citizens' Charter Unit (1996), *Handling Complaints Better. Proceedings of a Seminar to Promote Good Practice on the Passenger Rail Industry* (London, 19 March).

Centre d'Etudes sur les Réseaux, les Transports et l'Urbanisme (CERTU) et al. (2005), *Annual Survey on Urban Transport* (Paris: CERTU-DGMT-GART-UTP).

Centro Studi Federtrasporto (2001), *Bollettino economico sul settore dei trasporti, No 11. Il mercato del trasporto di passeggeri* (Rome: Author's Edition).

Coffey, A. (2002), 'Planung – durch Mitbestimmung Akzeptanz fördern', in Häupl, M. (ed.), *BürgerInnenbeteiligung und politische Partizipation. Konzepte zur Entwicklung der Demokratie in der Stadt* (Vienna: Promedia), pp. 104–20.

Community of European Railway and Infrastructure Companies (CER) (2005), *Implementation of the Charter on Rail Passenger Services in Europe. Progress Report September 2005* (Brussels: Author's Edition).

Consiglio Nazionale dell'Economia e del Lavoro (CNEL) (2005), *Trasporto pubblico locale – Osservazioni e proposte – Assemblea, 28 aprile 2005* (unpublished).

Czech, T., Middelberg, U. and Röhrleef, M. (2002), 'Dienstleistungsqualität im ÖPNV: Wie lässt sie sich messen?', *Der Nahverkehr* 9, pp. 25–30.

Czerwenka, B., Heidersdorf, C. and Schönbeck, M. (2001), *Eisenbahn-Beförderungsrecht, 4. Aufl.* (Berlin, loose-leaf-collection, August).

D'Alessandro, G. (2000), 'La Carta dei Servizi', presentation at the conference *La qualitá per la costruzione dei servizi pubblici locali*, 15 October, Parma.

Däubler, W. (2003), *Zugverspätungen als Rechtsproblem* (in NJW, 2651).

Dente, Bruno (1999), 'Sub-National Governments in the Long Italian Transition', *West European Politics* 20:1, pp. 176–93.

Deutsche Bahn AG (2005), *Mehr Rechte für die Fahrgäste, auch wenn dadurch die Bahn langsamer und teurer werden muss? Stellungnahme zur NRW-Bundesratsinitiative zur Stärkung der Fahrgastrechte.*

Deutsche Verkehrswissenschaftliche Gesellschaft eV (DVWG) (2001), *Kommunikation und Beteiligung bei Verkehrsprojekten – Beschleunigung oder Behinderung?* (Bergisch Gladbach: DVWG).

Deutschen Bank Research (2001), *Öffentlicher Personennahverkehr auf dem Weg in den Wettbewerb* (Frankfurt [Main]: Author's Edition).

Deutscher Bundestag (n.d.), *Stichwort Petitionen* (Berlin: Author's Edition).

Dick, M. (2002), 'Auf den Spuren der Motive, Auto zu fahren', *Verkehrszeichen* 4, pp. 9–16.

Dienel, P.C. (2002a), 'Im Dialog Beteiligungspotenziale nutzen. Innovation Planungszelle erschließt Perspektive', Häupl, M. (ed.), BürgerInnenbeteiligung und politische Partizipation. Konzepte zur *Entwicklung der Demokratie in der Stadt* (Vienna: Promedia), S. 50–61.

Dienel, P.C. (2002b), *Die Planungszelle. Eine Alternative zur Establishment-Demokratie*, 5th edn (Opladen: Westdt. Verl.).

Dienel. H.-L. (2005), *Webunterstützte Beteiligungsverfahren in der Verkehrsplanung* (1 February).

DIN EN 13816:2002, Transport – Logistik und Dienstleistungen – Öffentlicher Personenverkehr, Definition, Festlegung von Leistungszielen und Messung der Servicequalität.

DIN EN 13816:2002, Transport – Logistik und Dienstleistungen – Öffentlicher Personenverkehr, Definition, Festlegung von Leistungszielen und Messung der Servicequalität.

DIN EN 15140:2006, Öffentlicher Personennahverkehr – Grundlegende Anforderungen und Empfehlungen für Systeme zur Messung der erbrachten Dienstleistungsqualität.

DIN EN 15140:2006, Öffentlicher Personennahverkehr – Grundlegende Anforderungen und Empfehlungen für Systeme zur Messung der erbrachten Dienstleistungsqualität.

DIN EN ISO 9000:2005, Qualitätsmanagementsysteme – Grundlagen und Begriffe.

DIN EN ISO 9000:2005, Qualitätsmanagementsysteme – Grundlagen und Begriffe.

DIN EN ISO 9001:2000, Qualitätsmanagementsysteme – Anforderungen.

DIN EN ISO 9001:2000, Qualitätsmanagementsysteme – Anforderungen.

Direction des Transports Terrestres (DTT) (2003), *Urban Public Transport in France (Institutional Organization)* (Paris): http://www.certu.fr/, accessed June 2007.

Dostal, A. and Dostal, G. (1999), *Marktstudie des Reisebusverkehrs in Europa* (Bremerhaven: Berichte der Bundesanstalt für Straßenwesen: M, Mensch und Sicherheit 106).

Doyle, M. (2003), 'The Use of ADR in Ombudsman Processes': http://www.bioa.org.uk (accessed April 2004).

Drüke, H. (2004), 'Europas Stiefel drückt und zwickt – Grundprobleme der Wirtschaft Italiens', *Aus Politik und Zeitgeschichte* B 35-36, pp. 18–25.

Ebbers, L. (2005), 'Lobbyarbeit für den öffentlichen Verkehr – worauf kommt es an?', presentation at the conference *More Customers through More Participation?*, 29–30 September, Berlin: http://www. busrep.de/tagung.html (accessed April 2007).

Eichmann, V. (2005), 'Nahverkehrsplanung unter neuen Rahmenbedingungen – Erfahrungen aus dem Tellus-Projekt', presentation at the conference *More Customers through More Participation?*, 29–30 September, Berlin.

Ekoloff, T. and Fulton, D. (2005), 'BVG-Kundengarantie. Erfahrungen seit 1997', presentation at the conference *More Customers through More Participation?*, 29–30 September, Berlin.

Emmerson, G. (2005), 'Can Councils Learn from Anti-Road Protesters?', *Local Transport Today*, 29 September, pp. 20–21.

Emnid (2005), 'Zufriedenheit der ÖPNV-Kunden rückläufig': http://www.tns-emnid.com/pdf/presse-presseinformationen/2003/2003_07_15_TNS_Emnid_OEPNV.pdf (accessed 16 April 2005).

Endemann, W. (1886), *Das Recht der Eisenbahnen* (Leipzig).

Europäisches Verbraucherzentrum Deutschland (2004), *Jahresbericht 2004* (Kehl: Author's Edition).

Europäisches Verbraucherzentrum Deutschland (n.d.), *Schlichtung – Eine Anleitung für den schnellen Weg zum Recht* (Kehl: Author's Edition).

European Commission (2004), *EU Energy and Transport in Figures. Statistical Pocketbook 2004* (Luxemburg: Author's Edition).

European Commission (2004a), Eurobarometer 219: 'Consumers' Opinions on Services of General Interest', Summary (Luxemburg: Büro f amtliche Veröff): http://ec.europa.eu/public_opinion/archives/ebs/ebs_219_report_en.pdf.

European Commission (2005), *Strengthening Passenger Rights within the European Union* (Brussels, 16 February, KOM[2005] 46).

European Opinion Research Group (EORG) (2002), *Consumers' Opinion on Services of General Interest (Summary of the Report)* (Health and Consumer Protection Directorate-General: Eurobarometer 58, Special Edition).

EUSG-Konsortium (2006a), *Evaluation and Monitoring of Trends with Regard to Passenger Needs on the Level of Service and Treatment of Passengers. Final Report December 2006* (Berlin: unpublished).

EUSG-Konsortium (2006b), *Evaluation and Monitoring of Trends with Regard to Passenger Needs on the Level of Service and Treatment of Passengers. Annex: Country Report Italy* (Berlin [unpublished]).

Faas, H.-P. (n.d.), 'Qualitätstester der BVG' (unpublished manuscript).

Faivre d' Arcier, B. (2003), 'Urban Transport in France (Moving to a More Sustainable Policy)' (Transport Economics Laboratory): http://halshs.archives-ouvertes.fr/docs/00/06/96/21/PDF/BFA_Senshu_2003.pdf (accessed May 2007).

Falkemark, G. (1999), 'Is and Ought in Swedish Traffic Planning', in Andersson-Skog, L. and Krantz, O. (eds), *Institutions in the Transport and Communications Industries* (Canton, MA: Science History Publications), pp. 301–19.

Faulenbach, D. (2005), *Fahrgastrechte bei Zugverspätungen* (NZV, 398).

Faust, A. (2001), 'Sicherung von Akzeptanz – Rahmenbedingungen des Einsatzes von Planungszellen in der Verkehrsplanung', in Deutsche Verkehrswissenschaftliche Gesellschaft DVWG (ed.), *Kommunikation und Beteiligung bei Verkehrsprojekten* (Bergisch-Gladbach: DVWG), pp. 184–92.

FGSV (eds) (2004), *Verlässliche Bedienung im öffentlichen Personenverkehr. Empfehlungen zur Vermeidung von Verspätungen, Anschlussverlusten und deren Auswirkungen* (Köln: FGSV-Verlag [FGSV-Arbeitspapier No. 64]).

Finger, H.-J. and Eiermann, R. (1999), *Eisenbahntransportrecht* (loose-leaf-collection, January, Munich).

Fischer, C. et al. (2003), *Die Kunst, sich nicht über den Runden Tisch ziehen zu lassen. Ein Leitfaden für Bürgerinitiativen in Beteiligungsverfahren* (Bonn: Stiftung Mitarbeit).

Fischer, T.B. (1999), The Consideration of Sustainability Aspects within Transport Infrastructure Related Policies, Plans and Programmes, *Journal of Environmental Planning and Management* 42:2, pp. 189–219.

Fischer, Thomas B. (2006), 'Gestufte Verkehrsplanung für eine nachhaltige Entwicklung' (Institut für Regionalentwicklung und Strukturplanung): http://www.irs-net.de/download/workpaper2.pdf (accessed July 2006).

Fliegel, M. and Schiefelbusch, M. (2005), 'Passenger Advisory Boards – Concepts and Experiences in Germany and Austria', presentation at the conference *More Customers through More Participation?*, 29–30 September, Berlin.

Freise, R. (1999), 'Die Reform des internationalen Eisenbahn-Personenverkehrsrechts – CIV 1999', *Gedächtnisschrift für Johann Georg Helm* (Berlin: Duncker and Humblot), pp. 59 et seq.

Fürst, D., Scholles, F. and Sinning, H. (1998), 'Participative Planung: http://www-user.tu-chemnitz.de/~koring/quellen/paed01/kreativitaets-techniken/ptm_part.htm#bezug.

Gangemi, G. (2004), 'Regieren und Zivilgesellschaft in Zeiten der Regierung Berlusconi', *Aus Politik und Zeitgeschichte* B 35–6, pp. 39–46.

Gegner, M. (2002), *Die Wege des urbanen Verkehrs zur Daseinsvorsorge* (Berlin: Wissenschaftszentrum Berlin, WZB papers FS II 02 – 106).

Gesetz über den öffentlichen Personennahverkehr im Land Brandenburg (ÖPNVG), in der Fassung vom 26. October 1995.

Gesetz über die Planung, Organisation und Gestaltung des öffentlichen Personennahverkehrs (ÖPNVG) für das Land Baden-Württemberg, in der Fassung vom 8. Juni 1995, zuletzt geändert am 17. Juni 1997.

Giammetti, L. (2000), 'All'ascolto di chi viaggia, molteplica i numeri verdi', *Trasporti pubblici* 5, pp. 15–17.

Gillwald, K. (1997), *Ein Fall von Car Sharing: Umweltentlastung durch soziale Innovation* (Berlin: WZB, Abt. Sozialstruktur und Sozialberichterstattung).

Godwin, B.F. and Patterson, P.G. (1999), 'Consumer Coping Strategies with Dissatisfactory Service Encounters: A Preliminary Investigation', *Journal of Consumer Satisfaction, Dissatisfaction and Complaining Behavior* 12, pp. 145–54.

Grasse, A. (2004), /Italienische Verhältnisse 2004/, *Aus Politik und Zeitgeschichte* B 35–6, pp. 6–16.

Haas, H.-D. and Störmer, E. (1999), 'Angebotsqualität bei ÖV-Unternehmen', *Internationales Verkehrswesen* 51:4, pp. 119–24.

Haeske, U. (2001), *Beschwerden und Reklamationen managen (Kritische Kunden sind gute Kunden)* (Weinheim and Basel: Beltz Verlag).

Hanson, M. (1999), 'Exit or Voice: the Prospects for Public Transport User Representation in Israel', *World Transport Policy and Practice* 5:4, pp. 36–41.

Hart, C. (1988), 'The Power of Unconditional Service Guarantees', *Harvard Business Review* 66:4, pp. 54–62.

Häupl, M. (ed.) (2002), *BürgerInnenbeteiligung und politische Partizipation. Konzepte zur Entwicklung der Demokratie in der Stadt* (Vienna: Promedia).

Heger, M. (2003), 'Mediations-, Schlichtungs- und Schiedsverfahren in Deutschland', presentation at a seminar of the Clearingstelle Deutschland on 5 December, European Parliament, Straßburg.

Hesse, M. (1993), *Verkehrswende. Ökologisch-ökonomische Perspektiven für Stadt und Region* (Marburg: Metropolis).

Hilgers, M. (1994), 'Psychische Widerstände gegen eine verkehrspolitische Wende', *Universitas* 49, pp. 1055–65.

Hilgers, M. (1997), *Ozonloch und Saumagen. Motivationsfragen der Umweltpolitik* (Stuttgart/Leipzig: Hirzel).

Hilgers, M. (2001), 'Easy Going von Haustür zu Haustür – Psychologische Aspekte der Verkehrsmittelwahl', 21 – Das Magazin für zukunftsfähige Bildung, Edition 0/2001: http://www.oekom.de/verlag/german/periodika/21/text_0_2000.htm (accessed 17 May 2005).
Hilgers, M. (2003), 'On the Road Again – psychologische Aspekte innovativer Verkehrspolitik', presentation at the conference *König Kunde bei Bus und Bahn? – Verbraucherrechte im ÖPNV*, 12–13 June, Bonn.
Hinte, W. and Karas, F. (2003), 'Die Aktionsforschung in der Gemeinwesenarbeit', in Lüttringhaus, M. and Richers, H. (eds), *Handbuch Aktivierende Befragung. Konzepte, Erfahrungen, Tipps für die Praxis* (Bonn: Stiftung Mitarbeit), pp. 36–53.
Hippel, E. von (2001), 'Präventiver Verbraucherschutz: Vorbeugen ist besser als Heilen', *Aus Politik und Zeitgeschichte* B 24, pp. 16–22.
Hirschman, A.O. (1970), *Exit, Voice, and Loyalty* (Cambridge, MA: Harvard University Press).
Höfler, L. (2004), 'Feigenblatt oder wirksames Instrument für Fahrgastnutzen?', *Der Nahverkehr* 3, pp. 59–64.
Holst, F. von (2001), *Bürgerbeteiligung bei der Zulassung von Magnetschwebebahnprojekten, Zugleich eine Untersuchung der Regelung der Bürgerbeteiligung im Entwurf der Unabhängigen Sachverständigenkommission zum Umweltgesetzbuch (UGB-KomE)* (Aachen: Shaker).
Hopp, H. (1992), *Beauftragte in Politik und Verwaltung* (Bonn: Stiftung Mitarbeit).
Howlett, M. and Ramesh, M. (1995), *Studying Public Policy: Policy Cycles and Policy Subsystems* (Toronto: Oxford University Press).
Huefner, J.C. and Hunt, H.K. (1994), 'Extensions of the Hirschman Model: When Voice and Exit Don't Tell the Whole Story', *Journal of Consumer Satisfaction, Dissatisfaction and Complaining Behavior* 7, pp. 267–70.
infas/DIW (2004), *Mobilität in Deutschland 2002. Ergebnisbericht* (Berlin: Author's Edition)
Istituto per la promozione industriale (2003), *Dalla carta dei servizi alla carta della qualità* (Rome: Author's Edition).
Istituto Superiore di Ricerca e Formazione per i Trasporti (ISFORT) (2005), *Le politiche promosse in 10 città italiane* (Rome: Author's Edition).
Istituto Superiore di Ricerca e Formazione per i Trasporti (ISFORT) (2006a), *Il trasporto urbano su rotaia in Italia.* (Rome: Author's Edition).
Istituto Superiore di Ricerca e Formazione per i Trasporti (ISFORT) (2006b), *Lo stato di attuazione della riforma del tpl nelle regioni italiane* (Rome: Author's Edition).
Istituto Superiore di Ricerca e Formazione per i Trasporti (ISFORT) (2006c), *Dove vanno a finire i passeggeri? 3° Rapporto sulla mobilità urbana in Italia* (Rome: Author's Edition).
Jochim, H.E. (2007), 'Verspätungsentschädigungen der Bahnen', *Internationales Verkehrswesen* 59:3, pp. 95–9.

Johansson, T. (2007), 'Alternative Streitbeilegung – das Beispiel Schweden', presentation at the conference *Zukunft braucht Schlichtung – Schlichtung braucht Zukunft' of the Schlichtungsstelle Mobilität*, 29 March, Berlin.

Jungk, R. and Müllert, N.R. (1989), *Zukunftswerkstätten. Mit der Phantasie gegen Routine und Resignation* (München: Heyne).

Kamphausen, C. (1999), 'Das Marburger Verkehrsforum – Medium verkehrspolitischen Fortschritts?', *Verkehrszeichen* 4, pp. 27–32.

Karl, A. (2002), *Öffentlicher Verkehr im künftigen Wettbewerb – wie ein inkonsequenter Ordnungsrahmen und überholte Finanzstrukturen attraktive öffentliche Angebote verhinder* (Berlin: WZB discussion papers FS II 02-111).

Kelly, J. et al. (GUIDEMAPS-Konsortium) (2005), *Successful Transport Decision-making, A Project Management and Stakeholder Engagement Handbook – Practice Example Summaries* (Aachen/London: Author's Edition).

Kittel, T., Friebe, K. and Hay, E. (1928), *Die Eisenbahn-Verkehrsordnung*, Berlin.

Klein, A. (1996), 'Die französische Nahverkehrsabgabe', *Der Nahverkehr* 6, pp. 54–62.

Kleinfeld, R., Willems, U. and Zimmer, A. (2007), 'Lobbyismus und Verbändeforschung: Eine Einleitung', Kleinfeld, R., Willems, U. and Zimmer, A. (eds), *Lobbyismus* (Wiesbaden: Verlag für Sozialwissenschaften).

Klenke, D. (2007), 'Verkehrspolitiken', Schöller, O., Canzler, W. and Knie, A. (eds), *Handbuch Verkehrspolitik* (Wiesbaden: VS Verlag für Sozialwissenschaften).

Klotz, G. (2002), *Bahnbau und Bürgerbeteiligung* (Kassel).

Klühspies, J. (2001), 'Psychosoziale Grundlagen für ÖPNV-Konzepte', in Apel, D. et al. (eds), *Handbuch der kommunalen Verkehrsplanung* (Bonn: Economica).

Konrad, K. and Truffer, B. (2003), 'Möglichkeiten und Grenzen der Nutzerpartizipation in der Produkt- und Dienstleistungsentwicklung im öffentlichen Verkehr' (Kastanienbaum, unpublished).

Kuklinski, O. (2001), *Verkehrsforen in Hannover – Verkehrsplanung mit Beteiligung von Bürgerinnen und Bürgern, Erfahrungen und Schlussfolgerunegn aus einer Untersuchung von Stadtteilverkehrsforen und anderen Bürgerbeteiligungsverfahren zum Verker auf Stadtteilebene, Diskussionen Debatten Dialog* (Hannover: Bürgerbüro Stadtentwicklung).

Kuklinski, O. (2002), 'Erfahrungsaustausch von Akteuren aus Bürgerbeteiligungsverfahren zum Verkehr' (notes from a meeting on 25 October, unpublished).

Landwell & Associates et al. (2003), *Study on Penalty Systems in the Railway Industry* (Brussels: Report to European Commission, DG Energy and Transport, unpublished).

Langweg, A., Beckmann, K.J. et al. (2005), 'Emotionales Marketing im ÖPNV – ein Werkstattbericht aus dem Projekt "Lernen vom Pkw"', *Stadt Region Land* 80, pp. 29–38.

Larsson, S.O. and Göran, B. (2000), *Beslutsprocesser och planeringsresultat. En kunskapsöversikt* (Stockholm: Fritzes).

Laumann, G., Sauer, A. and Röhrleef, M. (2002), 'Servicegarantien. Ein (noch) neues Thema im öffentlichen Verkehr', *Internationales Verkehrswesen* 54:6, pp. 292–4.

Leborgne, G. (2005), 'Les lobbies du transport', *Rail and Transports* 23:2, pp. 24–33.

Legambiente (2006), *Primo Rapporto sul trasporto pendolare ferroviario nelle città italiane* (Rome: Author's Edition).

Lehmann, C. (2000), *Effiziente Koordination von Verkehrsleistungen im Öffentlichen Personennahverkehr* (Göttingen: Vandenhoeck und Ruprecht).

Ley, A. and Weitz, L. (eds) (2003), *Praxis Bürgerbeteiligung, Ein Methodenhandbuch* (Bonn: Stiftung Mitarbeit).

Linder, W. and Vatter, A. (1995), 'Kriterien zur Evaluation', Bischoff, A., Selle, K. and Sinning, H. (eds), *Informieren, Beteiligen, Kooperieren* (Dortmund: Dortmunder Vertrieg für Bau- und Planungsliteratur).

Lobmeyer, A. and Zandke-Schaffhäuser, B. (2007), 'Die Schlichtungsstelle Mobilität beim VCD', presentation at the conference *Zukunft braucht Schlichtung – Schlichtung braucht Zukunft' of the Schlichtungsstelle Mobilität*, 29 March, Berlin.

London Regional Passengers Committee (LRPC) (1996), *Submission to House of Commons Public Service Committee's inquiry into the Citizen's Charter programme* (London, October, unpublished).

London Regional Passengers Committee (LRPC) (2000), *Yours Disgusted Yours Delighted* (London: Case Studies in Complaint Handling. Author's Edition).

London Regional Passengers Committee; Central Rail Users Consultative Committee (LRPC/CRUCC) (1996), *Handling Complaints Better*, proceedings of a seminar to promote good practice on the passenger rail industry (London: Author's Edition).

London Transport Users Committee LTUC (2002), *London on the Move – Transport Policies for a Liveable City* (London: LTUC): http://www.ltuc.org.uk/view_document.php?id=391, accessed March 2004.

Lundsgaard-Hansen, N. et al. (2000), *Gesellschaftliche und politische Einflüsse auf Infrastruktur-Großprojekte* (Forschungsauftrag 58/97 auf Antrag des Bundesamtes für Strassen. Bern: Vereinigung Schweizerischer Strassenfachleute).

Lüttringhaus, M. (2003), 'Voraussetzungen für Aktivierung und Partizipation', in Lüttringhaus, M. and Richers, H. (eds), *Handbuch Aktivierende Befragung. Konzepte, Erfahrungen, Tipps für die Praxis* (Bonn: Stiftung Mitarbeit).

Lüttringhaus, M. and Richers, H. (2003), 'Einführung', in Lüttringhaus, M. and Richers, H. (eds), *Handbuch Aktivierende Befragung. Konzepte, Erfahrungen, Tipps für die Praxis* (Bonn: Stiftung Mitarbeit), pp. 6–8.

Luz, F. and Oppermann, B. (1993), 'Landschaftsplanung umsetzungsorientiert', *Garten und Landschaft* 11.

Maggi, S. (2005), *Storia dei trasporti in Italia* (Bologna: Il Mulino).

Maiwald, K.-O. (2004), 'Die Anforderungen mediatorischer Konfliktbearbeitung', *Zeitschrift für Rechtssoziologie* 25:2, pp. 175–89.

Meffert, H. (ed.) (2000), *Verkehrsdienstleistungsmarketing. Marktorientierte Unternehmensführung bei der Deutschen Bahn AG* (Wiesbaden: Gabler Verlag).

Meffert, H. and Bruhn, M. (2003), *Dienstleistungsmarketing (Grundlagen – Konzepte – Methoden)* 4th edn (Wiesbaden: Meffert-Marketing-Edition).

Mettler-v. Meibom, B. (1999), 'Kommunikation als strategische Planungsressource: http://www.perspektivenwerkstatt-essen.de/lese/artikel/mettler1.htm, accessed August 2003.

Meyer, A. and Blümelhuber, C. (1999), 'Kundenbindung durch Services', in Bruhn, M. and Homburg, C. (eds), *Handbuch Kundenbindungsmanagement* (Wiesbaden: Gabler Verlag), pp. 189–213.

Meyer-Liesenfeld, S. (1997), *Initiierung von Fahrgastbeiräten im Saarland* (Saarbrücken: VCD Landesverband Saarland [Projektbericht, 2. Auflage])

Meyer-Liesenfeld, S. (1997a), 'Initiierung von Fahrgastbeiräten im Saarland', *Verkehrszeichen* 1, pp. 19–22.

Michel, S. (2000), 'Mitwirkungsverfahren bei der Angebotsdefinition im öffentlichen Verkehr' (Bern: unpublished report for Amt für öffentlichen Verkehr).

Mietzsch, O. and Oellers, F. (2001), 'Die Rolle des Aufgabenträgers im liberalisierten ÖPNV-Markt', *Der Städtetag* 4, pp. 30–34.

Miller, N. (2005), 'Wofür sind Normen gut, warum brauchen wir sie?, presentation at the conference of railtec: http://www.railtec.de/files/ph_galerie/files/Kongress/Dr_Norbert_ Miller.pdf.

Ministerium für Wirtschaft und Mittelstand, Energie und Verkehr NRW (2001), *Masterplan Qualität* (Düsseldorf: Author's Edition).

Ministerium für Wirtschaft und Mittelstand, Energie und Verkehr NRW (eds) (2004), *Weichen stellen für zeitgemäße Fahrgastrechte, Verbraucherrechte im Öffentlichen Schienen-Personenverkehr* (Düsseldorf: Author's Edition).

Moritz, C. (2003), 'Außergerichtliche Streitbeilegung in Verbraucherrechtsstreitigkeiten', presentation to a seminar of the Clearingstelle Deutschland on 5 December, European Parliament, Strasbourg.

Müller, E. (2001), 'Grundlinien einer modernen Verbraucherpolitik', *Aus Politik und Zeitgeschichte* B 24.

Müller, G. (2004), 'Mobilität organisieren (Rahmenbedingungen für ein effektives Mobilitätsmanagement, *Internationales Verkehrswesen* 9, pp. 371–8.

Murphy, J.A. (1994), *Dienstleistungsqualität in der Praxis (Ein Handbuch für den praktischen Gebrauch)* (Munich and Vienna: Carl Hanser Verlag).

Mutz, G. (1975), 'Das Landesrecht nach Art. 54 CIM', *Zeitschrift für den internationalen Eisenbahnverkehr* 145.

Nennen, H.-U. (2001), 'Daseinsvorsorge: Zwischen Staat und Gesellschaft', in Steierwald, M. and Brenner, J. (eds), *Streitfragen im Verkehrswesen: Verkehr im Spannungsfeld zwischen Fortschritt und öffentlicher Alimentierung* (Stuttgart: TA-Akademie Baden-Württemberg, Arbeitsbericht 188), pp. 31–52.

Neugebauer, N. (2006), 'Servicegarantien im öffentlichen Verkehr in Deutschland. Vergleich bisheriger Konzepte und Erfahrungen', Diploma thesis, Humboldt-University of Berlin, Department of Transport Geography (unpublished).

Neugebauer, N. (2007), 'Servicegarantien im ÖPNV. Analyse der bisherigen Angebote auf dem deutschen Verkehrsmarkt und Potenziale der Optimierung', *Der Nahverkehr* 6, pp. 36–41.

Nicotera, A. and Santel, A. (1999), 'Italien: Der Reformprozeß im öffentlichen Nahverkehrswesen – Welche Chancen bietet er für die Verkehrsbetriebe?', *Public Transport International [German Edition]* 48:5, pp. 50–52.

Nordic Council of Ministers (2002), *A Study on ADR and Cross-border Complaints* (unpublished).

OPTEM and EU Commission, DG Health and Consumer Protection (2003), *European Consumers and Services of General Interest. Qualitative Study in the 15 Member States and the 10 Future Member States Acceding to the European Union in 2004* (Versailles/Brussels: Author's Edition): http:// ec.europa.eu/ public_opinion /quali/ql_sgi_report_en.pdf.

Osmotherly, E. (2003), 'Ombudsmen', *Adviser* 100, pp. 6–10.

Osservatorio Territoriale Infrastrutture Nordovest (2007), Rapporto 2006: http://www. otinordovest.it/pagine/documenti.asp (accessed June 2007).

Pfeiffer, M. (2001), 'Zufriedenheitsmessung durch Kundenbefragungen', *Der Nahverkehr* 1–2, pp. 48–55.

Pinardi, M. (1994), 'Stadterneuerung in Bologna', Bodenschatz, H. et al. (eds), *Stadterneuerung im Umbruch* (Berlin: TU, Institut für Stadt- und Regionalplanung), pp. 185–212.

Pinney, J.K. (1993), 'The Citizen's Charter: The United Kingdom's Commitment to Customer Satisfaction in the Public Service Sector', *Journal of Consumer Satisfaction, Dissatisfaction and Complaining Behavior* 6, pp. 141–57.

Pohar, M.A. (2004), *Der Vorschlag für eine Verordnung über die Rechte und Pflichten der Fahrgäste im grenzüberschreitenden Eisenbahnverkehr KOM (2004)* (143, endg., Reiserecht aktuell – RRa 2004, 194.

Pohar, M.A. (2004), *Schadensersatz wegen Informationspflichtverletzungen bei Zugausfall und Verspätung* (in NZV, 72).
Pohar, M.A. (2006), 'Rechtsbeziehungen zwischen Fahrgast und Eisenbahn – Vertragliche Pflichten und Haftung im nationalen und internationalen Eisenbahnverkehr in Deutschland und Europa' (Jena: zugl. dissertation, Münster).
Pohar, M.A. (2006), *Zum neuen § 17 EVO und dem Gesetzesvorhaben zur Stärkung der Fahrgastrechte* (VuR, 342).
Pohl, J. (1995), 'Italien dreigeteilt?', *Geographische Rundschau* 47:3, pp. 150–55.
Potthast, J. and Dienel, H.-L. (2003), *SMARTBENCH (Studie über Strategien, Programme und Projekte im Bereich der Verkehrsforschung ausgewählter europäischer Staaten, der USA und Japans, Teilstudie Frankreich)* (Berlin: unpublished).
Potz, P. (1998), '"Sostenibilità urbana"? Zur italienischen Debatte um Stadterneuerung und Nachhaltigkeit', *Jahrbuch Stadterneuerung 1998*, pp. 153–61.
Pro Bahn (2003), *Stellungnahme zur Anhörung des Bundestagsausschusses für Verbraucherschutz, Ernährung und Landwirtschaft vom 5.5.03* (ADrs 15 (10) 111: unpublished).
Pro Bahn (2004), *Verspätungen: Meistens hausgemacht* (and following articles),', Der Fahrgast 1/04,pp. 21et seq.).
Pro Bahn (2005), *Fahrgäste ohne Rechte: Ein gangbarer Weg zur Verbesserung. Stellungnahme zur NRW-Bundesratsinitiative zur Stärkung der Fahrgastrechte.*
Probst, G., Kahrs, C. and Rieschick-Dziabas, S. (2001), 'Was bindet Fahrgäste dauerhaft? Ereignisorientierte Kundenzufriedenheit im ÖPNV', *Der Nahverkehr* 3, pp. 8–12.
Projektgruppe Mobilität (2001), *Kurswechsel im öffentlichen Verkehr* (Berlin: Edition Sigma).
Prosperetti, L. et al. (2000), *Le Carte dei Servizi nella provincia di Milano* (Milan: Camera di Commercio).
Pulcher, J. and Kurth, S. (1996), 'Verkehrsverbund: The Success of Regional Public Transport in Germany, Austria and Switzerland', *Transport Policy* 2:4.
Rahs, R. (1996), 'Beteiligungsformen in öffentlichen Verkehrssystemen – europäische Beispiele im Vergleich', Bühler, T. (ed.), *Bürgerbeteiligung und lokale Klimaschutzpolitik*, proceedings of the conference *Bürgerbeteiligung als Motor der lokalen Klimaschutzpolitik* 7–8 December (Bonn: Wissenschaftsladen Bonn), pp. 61–70.
Rahs, R. and Hannover Region (1999), 'Expertenforum zur integrierten Kundenbeteiligung im ÖPNV der Hannover Region – Eine Vorbereitung auf den globalen Wettbewerb im Nahverkehrssektor' (issue record 29–30 September) (Hannover, unpublished).
Rajé, F. (2007), 'The Lived Experience of Transport Structure: An Exploration of Transport's Role in People's Lives', *Mobilities* 2:1 (March), pp. 51–74.

Regionalisierungsgesetz (RegG), in der Fassung vom 27. Dezember 1993 (BGBl I 1993, 2378, 2395).

Reichheld, F.F. (1997), 'Lernen Sie von abtrünnigen Kunden, was Sie falsch machen', *Harvard Business Manager*, pp. 57–68.

Reinert, A. (2003), 'Vorwort', in Lüttringhaus, M. and Richers, H. (eds), *Handbuch Aktivierende Befragung. Konzepte, Erfahrungen, Tipps für die Praxis* (Bonn: Stiftung Mitarbeit), p. 5.

Reinert, A. (2005), 'Bürger(innen)beteiligung: Dialog, Monolog oder wie? Aktuelle Ansätze und verschiedene Partizipationsmöglichkeiten', presentation at the conference *More Customers Through More Prticipation?*, 29–30 September, Berlin.

Reiß, C. (1985), *Rechtsgeschichte und Dogmatik der Inanspruchnahme von Personenbeförderungsleistungen der Eisenbahn* (Mannheim).

Reiter, H.-M. (1999), 'Verkehrskonzepte, Verkehrsplanung und Bürgerbeteiligung in Mittel- und Kleinstädten', Institut für Städtebau und Wohnungswesen der Dt. Akademie für Städtebau u Landesplanung: Congress *Verkehrsplanung für Städte und Regionen*, 18–19 October (Munich Author's Edition).

Rogge, L. (1999), Er*folgsbedingungen und Handlungsblockaden umweltorientierter kommunaler Verkehrspolitik (Handlungsansätze der Städte Hamm/Deutschland und Besançon/ Frankreich im Vergleich*) (Berlin: Schriftenreihe des IÖW 140/99).

Rui, S. (2004), 'Transport Policy and Public Involvement (Concertation between Mobilization and Frustration)', *The European Association: The European Journal of Social Science Research* 17:2, pp. 29–144.

Sabelberg, E. (1984), *Regionale Stadttypen in Italie* (Wiesbaden: Steiner [Geographische Zeitschrift]).

Säcker, R. and Rixecker, R. (eds) (2001), *Münchener Kommentar zum Bürgerlichen Gesetzbuch*, Vol. 1, *Allgemeiner Teil*, 4th edn (Munich), §§ 1–240.

Schäfer, B. (2004), 'Projektmanagement', Schophaus, M., Schön, S. and Dienel, H.-L. (eds), *Transdisziplinäres Kooperationsmanagement* (Munich: Ökom-Verl.), pp. 103–36.

Scheele, U. and Sterzel, D. (2000), *Öffentlicher Personennahverkehr zwischen Gemeinwohlinteressen und Markt* (Baden-Baden: Nomos Verl.-Ges.).

Schiefelbusch, M. (1995), 'Freizeitverkehre im ländlichen Raum', *Der Nahverkehr* 125.

Schiefelbusch, M. (1997), 'Stadtbahnfinanzierung in Großbritannien', *Der Nahverkehr* 12, pp. 62–7.

Schiefelbusch, M. (1998), 'Fahrgastvertretungen im britischen ÖV', *Der Nahverkehr* 3, pp. 60–65).

Schiefelbusch, M. (2005), 'Citizens' Involvement and the Representation of Passenger Interests in Public Transport Dimensions of a Long-neglected

Area of Transport Planning and Policy with Case Studies from Germany'. *Transport Reviews* 25:3, pp. 261–82.

Schiefelbusch, M. (2004), 'Kundenvertretung und Fahrgastrechte im öffentlichen Verkehr', Internationales Verkehrswesen 7+8, pp. 316–21.

Schiefelbusch, M. (2005), 'Öffentlicher Verkehr: "Beförderungsfälle" melden sich zu Wort', *Forschungsjournal Neue Soziale Bewegungen* 4, pp. 116–20.

Schiefelbusch, M. (2007), 'Reiseerlebnis und Verkehrswissenschaft – Stationen einer vorsichtigen Annäherung', in Dick, M. (ed.), *Mobilität als Tätigkeit: Die Perspektive der Akteure aus sozial- und verhaltenswissenschaftlicher Sicht* (Lengerich: Pabst).

Schiefelbusch, M. (2007a), 'Schlichtungsstellen und Ombudsleute im öffentlichen Verkehr – Schlichten, aber richtig!', *Der Nahverkehr* 11, pp. 50–55.

Schiefelbusch, M. (2007b), 'Paradigms of Transport Planning – the Rise (and Fall?) of the "Rational Transport User"' (presentation at 2nd Nordic Geographers' Meeting, 14–17 June, Bergen).

Schiefelbusch, M. and Bodensteiner, C. (2005), 'Bürgerbeteiligung bei der Netz- und Fahrplangestaltung des öffentlichen Verkehrs', in *Tagungsband zur AMUS* (Aachen: RWTH).

Schiefelbusch, M. and Fliegel, M. (2006), 'Fahrgastbeiräte als Ratgeber für Verkehrsunternehmen', *Der Nahverkehr* 7–8, pp. 48–53.

Schlichtungsstelle Mobilität (2005), 'Projekt "Schlichtungsstelle Mobilität"' (Erster Sachbericht mit Stand vom 31.03.2005) (Berlin, unpublished).

Schlichtungsstelle Nahverkehr (2003), *Projektbericht 2003* (Düsseldorf: Author' Edition).

Schmid, R. (1998), 'Marktbehauptung des ÖPNV – Sicht des Verbandes Baden-Württembergischer Omnibusunternehmer', in Herrmann, M. (ed.), *ÖPNV – Luxus oder Sparmodell?* (Stuttgart: Arbeitsbericht der Akademie für Technikfolgenabschätzung in Baden-Württemberg), pp. 87–96.

Schmidt-Bendun, R. (2004), *Internationales Eisenbahnverkehrsrecht made in Brüssel – Zum Verordnungsvorschlag der EG-Kommission KOM(2004)* (143 endg., GPR, 193).

Schmidtmann, S. (2006), *Neue Servicedienstleistungen für attraktive öffentliche Mobilitätsangebote in ländlichen Räumen Brandenburgs* (Arbeitspapier 11/2006) (Berlin: unpublished).

Schmithals, J. and Berkenhagen, J. (2004), 'Beziehungen und Motivation', in Schophaus, M., Schön, S. and Dienel, H.-L. (eds), *Transdisziplinäres Kooperationsmanagement* (Munich: Ökom-Verl.), pp. 69–102.

Schmithals, J., Schophaus, M. and Leder; S. (2004), 'Kooperationsmanagement in der Eventverkehrsplanung', in Dienel, H.-L. and Schmithals, J. (eds), *Handbuch Eventverkehr. Planung, Gestaltung, Arbeitshilfen* (Berlin: Erich Schmidt Verlag), pp. 130–45.

Schmucki, B. (2001), *Der Traum vom Verkehrsfluss – Städtische Verkehrsplanung seit 1945 im deutsch-deutschen Vergleich* (Frankfurt: Campus).

Schnieders, R. (2004), 'Die Regionalisierung des SPRVs in Frankreich', *Internationales Verkehrswesen* 6, pp. 256–8.

Schnippe, C. (2000), *Psychologische Aspekte der Kundenorientierung (Die Kundenzufriedenheit mit der Qualität von Dienstleistungen am Beispiel des ÖPNV)* (Frankfurt/Main: Peter Lang, Europäischer Verlag der Wissenschaften).

Schophaus, M. and Dienel, H.-L. (2003a), 'Bürgerausstellung – ein neues Beteiligungsverfahren für die Stadtplanung', *Forschungsjournal Neue Soziale Bewegungen* 15:2, pp. 90–96.

Schophaus, M. and Dienel, H.-L. (2003b), 'Die Bürgerausstellung', in Ley, A. and Weitz, L. (eds), *Praxis Bürgerbeteiligung. Ein Methodenhandbuch* (Bonn: Stiftung Mitarbeit), pp. 83–90.

Schophaus, M., Schön, S. and Dienel, H.-L. (eds) (2004), *Transdisziplinäres Kooperationsmanagement. Neue Wege der Zusammenarbeit zwischen Wissenschaft und Gesellschaft* (Munich: ökom).

Schröder, C. (2003), 'Kommunikative Verfahren in der Stadt- und Verkehrsplanung' (Berlin, unpublished).

Schüler, R. (2005), 'Wie finden Planer und Bürger den konstruktiven Weg? Erfahrungen aus der Mediation', presentation at AMUS, 26 July, Aachen.

Schuster, A. (1996), 'Widerstände bei der Umsetzung von Verkehrskonzepten – Ursachen und Handlungserfordernisse', Apel, D. et al. (eds), *Handbuch der kommunalen Verkehrsplanung*, Teil 3.2.9.2 (Heidelberg: Hüthig).

Seippel, A. (2003), 'Aktionsuntersuchung', in Lüttringhaus, M. and Richers, H. (eds), *Handbuch Aktivierende Befragung. Konzepte, Erfahrungen, Tipps für die Praxis* (Bonn: Stiftung Mitarbeit), pp. 25–32.

Selle, K. (2000), *Was? Wer? Wie? Warum? Voraussetzungen und Möglichkeiten einer nachhaltigen Kommunikation* (Dortmund).

Siefke, A. (1998), *Zufriedenheit mit Dienstleistungen (ein phasenorientierter Ansatz zur Operationalisierung und Erklärung der Kundenzufriedenheit im Verkehrsbereich auf empirischer Basis)* (Frankfurt/Main: Peter Lang, Europäischer Verlag der Wissenschaften).

Simpson, B.J. (1996), 'Deregulation and Privatization: The British Local Bus Industry following the Transport Act 1985', *Transport Reviews* 16, pp. 213–23.

Staudinger, A. (1999), *Verspätungsschaden nach Eisenbahnverkehrsordnung: Ein europarechtswidriger Anachronismus?* (NJW, 3664).

Staudinger, A. (2004), *Verbraucherrechte im Öffentlichen Schienen-Personenverkehr* (Münster: unpublished).

Staudinger, A. and Schiefelbusch, M. (2004), 'Fahrgastrechte und Kundeninteressen im öffentlichen Verkehr', presentation at *Kolloquium sozialwissenschaftliche Mobilitätsforschung* of the Wissenschaftszentrum, 8 July, Berlin.

Staudinger, J. von (2002), *Kommentar zum Bürgerlichen Gesetzbuch mit Einführungsgesetz und Nebengesetzen* (Berlin, EGBGB/IPR, 13. Bearbeitung).

Stauss, B. and Seidel, W. (2002), *Beschwerdemanagement (Kundenbeziehungen erfolgreich managen durch Customer Care)* 3rd edm (Vienna: Carl Hanser Verlag Munich).

Stiftung M. (ed.) (1996), *Bürgergutachten ÜSTRA. Zusammenfassung der Aussagen und Empfehlungen* (Bonn: Author's Edition).

Szynka, P. (2003), 'Aktivierende Gespräche bei Saul Davis Alinsky', in Lüttringhaus, M., Richers, H. (eds), *Handbuch Aktivierende Befragung. Konzepte, Erfahrungen, Tipps für die Praxis* (Bonn: Stiftung Mitarbeit), pp. 11–17.

Tänzler, D., Schulz, A. et al. (2005), 'Nachhaltigkeit und Verbraucherpolitik: Ansätze und Maßnahmen in anderen Ländern. Studie im Auftrag der Bundesanstalt für Landwirtschaft und Ernährung' (Berlin, unpublished).

Tavakoli, A.Al. (2001), *Privatisierung und Haftung der Eisenbahn* (Baden-Baden: Nomas Verl.-Ges.).

TNS Opinion and Social and European Commission DG SANCO (2005), Eurobarometer 'Passengers' Rights' (Special Eurobarometer 228), *Summary*: http://ec.europa.eu/ public_opinion/archives/ebs/ ebs_228_sum_en.pdf, accessed December 2007.

TNS Opinion and Social and European Commission DG MARKT (2005), 'Prices and Quality of Services of General Interest' (Special Eurobarometer 226 – Wave 53.1 Summer 2003): http://ec. europa.eu/public_opinion/archives/ebs/ ebs_226_en.pdf, accessed December 2007.

Töllner, M. (2002), *Methoden und Werkzeuge der Informellen Planung (ein Vergleich zwischen Deutschland und Frankreich)* (Aachen: Shaker).

Tonner, K. (2004), *Die EG-Verordnung über Ausgleichsleistungen bei Nichtbeförderung, Annullierung und großer Verspätung – ein wichtiger Beitrag des Rechts zur Qualitätssicherung im Luftverkehr* (Reiserecht aktuell- RRa, 52).

Topp, H.H. (2004), 'Über den Versorgungsauftrag hinaus ("König Kunde" hat hohe Ansprüche)', *Verkehr und Technik* 11.

Topp, H.H. (2006), 'Perspektiven des Nahverkehrs in einer wachsenden Metropolregion', *Der Nahverkehr* 1–2, pp. 21–6.

Varnier, C., Assouline, G. and Joly, P.-B. (2001), *Assessing Debate and Participative Technology Assessment (The French National Report)* (Annex 2 of EU Research Project ADAPTA, unpublished).

Vassalli, S. (1998), *Gli italiani sono gli altr,* (Milan: Baldini & Castaldi).

Verband Deutscher Verkehrsunternehmen (VDV) and Socialdata (1991), *Mobilität in Deutschland* (Köln: Author's Edition).

Verbraucherzentrale NRW (2003), *Schriftkommunikation der Deutschen Bahn AG (Eine Analyse der Schlichtungsstelle Nahverkehr der Verbraucherzentrale*

NRW Düsseldorf: http://www. schlichtungsstelle-nahverkehr.de/mediabig/35A.pdf, accessed March 2007.

Verbraucherzentrale NRW (2004), Geld terug bij vertraging ('Money Back in Cases of Delay'). Compensation of the Dutch Railways: http://www. schlichtungsstelle-nahverkehr.de/ mediabig/118A.pdf.

Verdoold, H. (2005), 'OV-platforms – Working Together for Public Transport', presentation at the conference *More Customers Through More Participation?*, 29–30 September, Berlin.

Viegas, J.M. and Macario, R. (2000), 'Aumento di qualitá nei sistemi di mobilitá urbana', *Trasporti pubblici* 5, pp. 29–36.

vorm Walde, R. (1997), 'Wie zufrieden sind die Fahrgäste?', *Der Nahverkehr* 5, pp. 63–5.

Wachholz, C. (2001), *Die Zukunftskonferenz – ein Planungs- und Dialoginstrument als Katalysator für die Ausgestaltung eins zukunftsfähigen Öffentlichen Personen-Nahverkehrs? Evaluation eines Fallbeispiels. Diplomarbeit im Studiengang Umweltwissenschaften an der Universität Lüneburg* (Lüneburg: unpublished).

Webler, T. and Renn, O. (1995), 'A Brief Primer on Participation: Philosophy and Practice', in *Fairness and Competence in Citizen Participation: Evaluating Models for Environmental Discourse* (Dordrecht: Kluwer).

Weirauch, W. and Heinze, W. (1962), *Eisenbahn-Verkehrsordnung*, 8th edn (Berlin: de Gruyter).

Weisbord, M.R. and Janoff, S. (1995), *Future Search. An Action Guide to Finding Common Ground in Organizations and Communities* (San Francisco: Berrett-Koehler).

Weiß, H.-J. (2003), 'Die Doppelrolle der Kommunen im ÖPNV', *Internationales Verkehrswesen* 7+8, pp. 338–42.

White, P.R (1995), 'Long-Run Impacts of Express Coach Deregulation in Britain', *Coach and Bus Week* 30 September, pp. 24–5.

White, P.R (1997), 'What Conclusions Can be Drawn about Bus Deregulation in Britain?', *Transport Reviews* 1, p. 11.

Wimmer, F. and Roleff, R. (1998), 'Beschwerdepolitik als Instrument des Dienstleistungsmanagements', in Bruhn, M. and Meffert, H. (eds), *Handbuch Dienstleistungsmanagement. Von der strategischen Konzeption zur praktischen Umsetzung* (Wiesbaden: Gabler Verlag), pp. 266–85.

Wirtz, J. (1998), Dienstleistungsgarantien als wirksames Mittel, um bessere Servicequalität zu erreichen, zu erhalten und zu vermarkten', in Meyer, A. (eds), *Handbuch Dienstleistungsmarketing*, Vol. 1 (Stuttgart: Schaeffer-Pöschel Verlag), pp. 828–45.

Zachcial, M. (1985), 'Das Ende eines Monopols. Eisenbahn und Kraftverkehr', in Eisenbahnjahr-Ausstellungs-GmbH (ed.), *Zug der Zeit – Zeit der Züge* (Berlin: Siedler), pp. 663–71.

Index